Mirror Matter

Pioneering Antimatter Physics

The Wiley Science Editions

Mirror Matter

Pioneering Antimatter Physics

Robert L. Forward, Ph.D.
and
Joel Davis

Wiley Science Editions

John Wiley & Sons, Inc.

New York • Chichester • Brisbane • Toronto • Singapore

For Paul A.M. Dirac
and Gene Roddenberry,
two creators of the future

Publisher: Stephen Kippur
Editor: David Sobel
Managing Editor: Andrew Hoffer
Editing, design, and production: Publishing Synthesis, Ltd.

This publication is designed to provide accurate and authoritative information in regard to the subject matter covered. It is sold with the understanding that the publisher is not engaged in rendering legal, accounting, or other professional service. If legal advice or other expert assistance is required, the services of a competent professional person should be sought. FROM A DECLARATION OF PRINCIPLES JOINTLY ADOPTED BY A COMMITTEE OF THE AMERICAN BAR ASSOCIATION AND A COMMITTEE OF PUBLISHERS.

Library of Congress Cataloging-in-Publication Data
Forward, Robert L.
     Mirror matter : pioneering antimatter physics / Robert L. Forward,
Joel Davis.
        p.   cm.—(Wiley science editions)
     Bibliography: p.249
     Includes index.
     ISBN 0-471-62812-3
     1. Antimatter. I. Davis, Joel, 1948– . II. Title.
     III. Series.
     QC173.F67 1988
     530—dc19   87-33020
                                                                           CIP

Printed in the United States of America

88 89 10 9 8 7 6 5 4 3 2 1

Preface

"Space: the final frontier . . . " These are the opening words to what is probably the most famous set of science fiction stories today. It is also the opening to a future in which antimatter is commonplace. The universe of *Star Trek* is a universe in which star-traveling spaceships use the enormous energy of matter-antimatter annihilation to leap across the galaxy at velocities much greater than light itself. And that's why it is science fiction. For nothing can travel faster than light.

But antimatter? And matter-antimatter annihilation? No: they are not science fiction. They are science fact.

Interspersed throughout this book you will find a continuing science fiction short story, showing how antimatter may be used in the future. Antimatter itself, though, is quite real. Antimatter-powered spaceships are not yet a reality, but the stuff to power them is as real as the paper on which these words are written. In the chapters to follow we take a look at what antimatter really is; how it was discovered; how it is made and stored, even today, by physicists around the world; how it may someday be made in quantities millions of times greater than today; and how, perhaps in the lifetimes of some people reading this book, it may be used to send spaceships to the stars.

Acknowledgments

Writing this book would not have been possible without the help and support of many people, including (but not limited to): Poul Anderson, Dr. Isaac Asimov, Dr. Bruno Augenstein, Anne Baittinger, Greg Bear, Dr. Roy Billinge, Frank Catalano, Marie Celestre, Dr. Owen Chamberlain, Dr. John Cramer, Hall Daily, Robert Finn, Henry Fuhrmann, Robert Heinlein, Dr. Steven Howe, Dr. Michael Hynes, David Morgan, Jr., Larry Niven, USAF Maj. Gerald Nordley, Dr. Jim Powell, USAF Capt. William Sowell, Dr. Hiroshi Takahashi, Jack Williamson, and Lynn Yarris. Any errors of fact in this book are not theirs, but ours. Also, any opinions we express are ours, and should not be construed in any fashion as those of any individual or organization mentioned in this book.

We also wish to thank our agent, Russell Galen, for getting us together; our editor, David Sobel, for putting up with us; and you, for reading our book.

Contents

Turn Left at the Moon:
I

E ric took yet another look at the mauve polka-dot clock with the bright green numbers. "1740 Friday 25 Sept 2035" blinked at him from the upper right corner of the screen. He still had another 20 minutes with the homework tutor before he could quit for the day. "The timer must be broken," he grumbled to himself, and changed the clock to red stripes with electric blue numerals.

The tutor was getting annoyed with Eric's slow response to the question.

"Come on, Eric," it said. "If the formula stays on much longer, you'll have phosphor burn-in."

The formula began blinking. "IF $6x+3=27$, WHAT IS x?"

Eric mumbled briefly to himself, typed "4", and leaned back in his chair. The screen erupted in a swirl of colors. "Excellent!" said the tutor. "Just 18 minutes left, Eric. You are way behind the time line! Speed it up! Here's the next one."

The fifteen-year-old sighed and leaned forward. Might as well, he thought to himself. His heart wasn't in it, but at least his parents would not be too upset if he finished before they got home from work. For a moment he thought about running an interrupt through the system and checking for any news on his Grandma Ginny— but then he remembered that the tutor AI program locked the terminal cold. The only way to break it was with a wormprog. And if he did that he'd be in deep guck.

—And then the clock chimed 1800. "Thank you, Eric," said the tutor. "Your score today is 35 dash 63 dash 4. Quite good, considering your anxiety level. Hang in and hang on, kid. Interface you tomorrow."—Click—. The AI signed off and shut down. Eric heard the faint high-pitched whistle of the 1810 Italian supertrain: his mother coming home from work.

He stood and walked over to the picture window that looked down from their home in Les Houches in the French Alps into the deep valley below. Now Eric could see the wheel-train from Switzerland off to the right. It was late tonight. His mother would beat his father to the station for the second time that week.

The supertrain coming in from Italy slowed from its 200 kph speed by pumping its kinetic energy back through the linear motor in the super-conducting levitracks, and came to a whispering halt at the station. The trees partially blocked his view, but soon he could see a tiny figure rapidly walking to a car parked at the far end of the lot. His mother climbed into their Nissan electric and waited for the wheeltrain from Geneva to arrive.

Eric turned from the window and gazed vacantly past the clutter of his room, and through the open door. The oven had turned on and he could smell dinner cooking. He sniffed. Chicken Kiev. A vivid memory swept through him. He had been, oh, maybe two years old. His parents kissing. He remembered how he had laughed. *Pressing mouths together. How—weird!* His own mouth twisted into a wry smile at the memory. His parents, he knew, had met right down there, in that station beyond the window and the trees and the tumbling slopes, in a crowd of holiday skiers. Kirsten Docherty had been a materials engineer from the Republic of California. James Davis was a physicist at the CERN nuclear research center near Geneva. Within two weeks she had decided to spend the rest of her life in the Alps, and he had decided that was

a great idea. Would she spend it with him? Yes. *Pressing mouths together* . . . Eric had seen the wedding pictures—a quiet Episcopal rite ceremony with his Aunt Joan presiding. His mother easily found a job with the European Space Agency at their rocket fabrication and test facility in Turin. Eric had been born a year after the marriage, and they had moved from his father's old apartment in Geneva to Les Houches shortly after that. Fifteen years ago now.

And in fifteen more years, where would he be?

His father's wheeltrain finally arrived, and soon the family's high-powered superconducting electric Nissan was zooming up the mountain roads. Eric started worrying about his grandmother again. She had been feeling very poorly of late. He was wrestling with the terrifying thought that he might never see her again. It didn't seem possible; all adolescents are self-imaged immortals. What was the point of it all? he thought. Who cares if $x=4$? Ginny was neat. Neat old people weren't supposed to . . . to . . .

He caught a glimpse of the car coming around the last hairpin curve and into the driveway. Suddenly Eric remembered the very interesting piece of gossip Michelle had passed on to him that morning on AnarkNet from the arbitrageur from Mexico City. Was Boeing/Mitsubishi really building a human-crewed interstellar torchship? His parents would find that rather interesting.

They did not find the gossip from the Mexican arbitrageur very interesting. They found Eric's presence on

AnarkNet irritating, and his home-work scores disturbing. Eric was told to clean up his room ("that pit," as his father called it) before he came down for dinner.

By then, the two adults had for-gotten their pique with him. "Good day at CERN," his father remarked over the chicken. "We had over 200 scientists in for the triple anniversary meeting. About a dozen Nobels among 'em."

"What was the anniversary?" Eric asked.

"Don't talk with your mouth full," said his father, his mouth full of rice. He swallowed. "Oops. Sorry. It's been 50 years since Rubbia and van der Meer got the Nobel Prize for finding the W and Z particles. They held the ceremony in the center of the old LEAR—which, by the way, the Swiss government has just declared a na-tional monument! Pretty good, huh?"

Eric's mother sipped her wine. "Well, it'll keep the accelerator engi-neers from cannibalizing it for any more spare parts," she said.

"Were they there?" asked Eric.

"Who?" said his dad.

"Those guys. Rubbia and, uh, what's-his-face."

"Oh. Well, van der Meer's been dead for years, and Rubbia's so busy being king of the hill on the moon with LUPEC that he decided not to come. At least, that's what Gabrielse said in his introductory remarks. Actually, the old man couldn't have come even if he wanted to. The gravity would kill him.

"Anyway," he continued, pausing for a moment to take another bite of chicken, "the second part of it is the 80th anniversary of the discovery of the antiproton at Berkeley, and the third part is the 102d anniversary of the discovery of the positron in 1933 at Caltech."

"I thought that was in 1932," said Eric's mother.

"Depends on whose history book you read. Bellefeuille-Rice says '32, Patterson says '33. I think—"

"Anniversaries are fun," his mother interrupted, "but time-consuming. Did you get anything more done on your carbon-cycle work?"

Eric wondered what his dad had been about to say about history.

"—frustrating. The antimatter en-gineers at the heavy ion collider factories are finding us enough anti-helium nuclei to keep our experiment supplied, but we still can't get the damned solar carbon cycle mirrored with three antihelium fused into anticarbon."

"But wasn't that the whole point? If the cycle doesn't work with anti-matter the same as it does with matter, then that might explain why we don't see antimatter anywhere in the universe."

"Well, maybe. Or it could be that we're just not doing the experiment correctly. The fusion machine works fine turning regular helium into car-bon. But I'm beginning to suspect that the few residual molecules of air in the vacuum chamber are somehow helping that reaction along. Then, when we try the same thing with antihelium, those residual molecules are of no help at all. They hinder the antimatter version, because they're normal matter, not antimatter. So, we'll try some variations and see what happens. Six months to a year is my guess on this one. In fact, we shut down for three months starting next week so the engineers can rebuild the machine with a new antiproton beam vacuum scrubber. We'll use it to annihilate any leftover air molecules. Get a better vacuum. Maybe that'll change things.

"What's up at ESA today? Blow any more engines?"

Eric snickered. His dad would not let his mother forget about that little faux pas.

She raised her eyebrows at her husband. "Actually, I got almost nothing done," she said. "I kept worrying about mother. Oh, I had the techs run a few of the new high temperature plasma rocket nozzles through the p-bar flaw detection facility, but I couldn't get up the interest to pull down the data to analyze the results."

She smiled ruefully and shook her head. "I spent most of the day re-arranging my computer files." She turned to Eric and ruffled his hair. "Sorry for snapping at you about your homework, Red," she said. "You're worrying too, aren't you."

"Well . . . yeah . . . " Eric replied.

"I'm sure Ginger'll be just fine," said his father. "But, really, staying under pressure 24 hours a day for weeks at a time is bound to cause ear problems."

"Mother's been an undersea environmentalist for nearly 30 years, Jim, and this is the first time ear problems have made her so dizzy she can't even swim, much less walk! I think this is serious."

Chapter One

In the Beginning

What's in a Name?

Say the word "antimatter" to most people, and their eyes begin to glaze over. It's as if you've said "quantum mechanics," or "theory of relativity," or "certificates of deposit." In fact, the concept of antimatter is quite simple—much simpler than certificates of deposit, or even relativity.

Antimatter is, put simply, the mirror image of matter. A mirror image of anything is easy to recognize and understand, and only a moment's thought is needed to distinguish it from the real thing. The left and right sides of the image are reversed from the original. Other than that, the image and the original look the same. The shape of your image in the mirror is the same as that of the original. If you turn a somersault, so will your mirror image, either forward or backward, identical to your somersault. Only left and right are reversed.

A similar situation exists for antimatter versus matter. You are made of normal matter. The normal matter is in the form of atoms. These include atoms of carbon, oxygen, hydrogen, iron, and many other elements. All of these atoms are made up of three different building blocks: protons, neutrons and electrons. The proton is heavy and has a positive electrical charge. The neutron is slightly more massive than a proton, and has no electrical charge. The proton and neutron combine together to form the positively charged nucleus of an atom. Nearly all of an atom's mass resides in the nucleus. Orbiting around the nucleus are one or more very light subatomic particles. These are the electrons, which have 1/1,836 the mass of a proton and a negative electrical charge. (Some basic abbreviations, constants, and scientific terms can be found in the Appendix.)

Physicists call the electron a "fundamental particle." As far as anyone has ever been able to determine, the electron is not made of anything smaller. It is a subatomic particle which is itself basic, fundamental. The electron is a type of fundamental particle called a *lepton*. The word comes from a Greek word meaning "small." Six lepton particles are known to exist: the electron, electron neutrino, muon, muon neutrino, tau, and tau neutrino, (Table 1–1).

Table 1-1   The Lepton Family of Fundamental Particles

Name	Lifetime	Discovered
Electron	Stable	1895
Electron neutrino	Stable	(1930) 1956
Muon	2.2×10^{-6} sec	1935
Muon neutrino	Stable	1962
Tau	$<2.3 \times 10^{-12}$ sec	1975
Tau neutrino	Stable	1975

Leptons are subatomic particles that have no internal structure (as far as we can tell today). They can be treated as point-like. Leptons interact with matter by means of the gravitational and weak nuclear forces. If they carry an electrical charge, like electrons, they also interact via the electromagnetic force. Leptons do not interact with other matter through the strong nuclear force. Neutrinos are electrically neutral and apparently have no mass. Therefore they interact with other matter only through the weak nuclear force.

Protons and neutrons, on the other hand, are themselves composed of still tinier entities called quarks (Table 1-2). There seem to be six brands of quarks. They come in various versions, with different characteristics like strangeness, charm, and color. Quarks, like electrons, are fundamental particles. They seem to be made of nothing smaller, and they combine in twos and threes to form the bewildering array of subatomic particles that physicists have discovered or made in their giant particle accelerators (Table 1-3).

Physicists have found that subatomic particles like electrons, neutrons, and protons have distinct characteristics. They're like fruits. Fruits might be said to have the characteristics of shape, color, and taste. Different fruits can be identified by different values for those characteristics. Oranges have one set of values for shape, color, and taste, and apples have another. Subatomic particles also have sets of characteristics. In addition to mass and charge, they have a characteristic called spin, as if they were spinning about an axis like a planet. Since the proton and electron have an electrical charge, their spin motion is the same as the rotation of an electrical charge in a circle. A moving electrical charge creates an electrical current. An electrical current creates a magnetic field. The proton and electron both have magnetic fields. (The Earth is spinning and has a magnetic field, but this field is caused by an internal dynamo effect that involves the Earth's mantle and core. It is not because the Earth has an electrical charge.)

Since the electron has an electrical charge and is spinning, that spinning charge produces a magnetic field. So the electron has a south magnetic pole coming out at its spin pole axis. The magnetic field gives the electron something called a magnetic moment and allows it to interact magnetically with other particles and magnetic fields.

What, then, would an "antielectron" be like? It turns out that the anti-electron has the same mass and spin of the electron. However, there is one

Table 1-2 Quarks and Some of Their Properties

Flavor	Charge (1)	Color (2)	Spin	Charm (3)	Strangeness (3)
Down	$-1/3$	r,g,b	1/2	0	0
Up	$+2/3$	r,g,b	1/2	0	0
Strange	$-1/3$	r,g,b	1/2	0	$+1$
Charmed	$+2/3$	r,g,b	1/2	$+1$	0
Bottom	$-1/3$	r,g,b	1/2	0	0
Top (4)	$+2/3$	r,g,b	1/2	0	0

All subatomic particles that are not leptons are made of some combination of two or three quarks. Such "non-elementary" elementary particles are called *hadrons*. They include the proton and the neutron, the two particles which compose the nucleus of all atoms. Hadrons are made of entities called *quarks*. The existence of quarks was first proposed in 1964 independently by Murray Gell-Mann and George Zweig of the California Institute of Technology. Each quark has its corresponding *antiquark*. Quarks and antiquarks have never been directly seen. Apparently they can only exist in combination with another quark, and can never be seen alone.

(1) For antiquarks the electrical charge property is reversed. Up, charmed, and top antiquarks have a charge of $-2/3$. Down, strange, and bottom antiquarks have a charge of $+1/3$.

(2) "Color" is not really color. It is a charge in the sense that the electrical charge is a charge. The three possible color charges for quarks are called red, green, or blue (r,g,b). Each of the six quarks comes in three varieties or colors. The color charge for antiquarks is reversed from that of quarks. That is, antiquarks have anticolors—antired, antigreen, and antiblue.

(3) "Strangeness" and "charm" are properties of quarks that cannot easily be compared to anything in our normal experience. Only the charm quark has charm, and only the strange quark has strangeness.

(4) The top quark has not yet been detected, but its existence is assumed on the basis of symmetry.

Table 1-3 A Few Quark Combinations

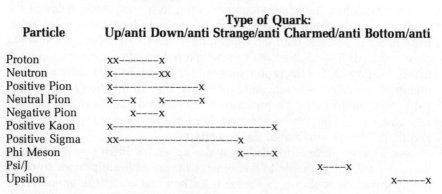

```
                                    Type of Quark:
      Particle      Up/anti Down/anti Strange/anti Charmed/anti Bottom/anti

Proton             xx--------x
Neutron            x---------xx
Positive Pion      x----------------x
Neutral Pion       x---x   x-------x
Negative Pion          x----x
Positive Kaon      x--------------------------------x
Positive Sigma     xx-----------------------x
Phi Meson               x-----x
Psi/J                            x----x
Upsilon                                  x-----x
```

Explanation: The proton, for example, is made of two up quarks (the two x's under "Up") and one down quark (the x under "Down") bound together (the dotted line connecting the three x's).

characteristic that is different between matter and antimatter. The antielectron has a *positive* electrical charge, of +1. Like your image in a mirror, the spin and "size" of the electron and antielectron are identical. The electrical charge, however, is opposite, just as left and right are reversed for you and your mirror image. Physicists call the antielectron a *positron* because it has a positive electrical charge. As shown in Figure 1-1, if we view the electron in a magic mirror that changes negative charge into positive charge, the positron is the mirror image of the electron. In the mirror, the left-to-right rotation of the electron is changed to a right-to-left rotation in the mirror. This reversal of handedness is called *parity reversal*. But since the charge has also been reversed, the magnetic fields of the electron and the positron have the same orientation.

The same is true of the proton and its mirror twin, the *antiproton*. The positive charge of the proton is mirrored as a negative charge in the antiproton and the spin direction is also reversed, giving similar orientation of the magnetic field. In both cases, however, the relationship of the direction of spin to the direction of the magnetic field is opposite for the normal matter particle and the antimatter particle. So the magnetic moment, in which the magnetic field direction is measured with respect to the spin axis orientation, is reversed for the antiparticles.

The same is true for the case of the *neutron* and *antineutron*, even though neither has a net electrical charge. Remember: the *neutron* (like the proton) is a complex object. It is made of three quarks. Quarks carry electrical charges that are a fraction of the charge of a proton. They have either a +2/3 charge or a –1/3 charge. In the neutron, two of those quarks each have a –1/3 electric charge. The third quark has a +2/3 electric charge. The result is obviously no electrical charge for the neutron; the three fractional charges of the quarks cancel out. For every quark, though, there is an antiquark. Those antiquarks have either a –2/3 charge or a +1/3 charge—mirror images of quarks. So an *antineutron* is made of two quarks with +1/3 charge and one with –2/3 charge. The result: an antineutron with no electrical charge.

This canceling out effect does not apply to the magnetic fields of the quarks, however. In the neutron, the magnetic field from the rotating negative quarks is stronger than the cancelling magnetic field from the rotating positive quark. The neutron ends up with a net negative magnetic moment that is roughly –2/3 of the proton magnetic moment. So the neutron and antineutron have opposite magnetic moments. The antineutron's north magnetic pole would be in the location of the neutron's south magnetic pole, and vice versa. That's how neutrons can have antineutron mirror images, and that's how physicists tell the two apart.

Since every particle needed to make up atoms has its antiparticle, it is conceivable in principle to combine positrons with antiprotons to make antihydrogen, as is shown in Figure 1–1. Then one could use antineutrons to make heavier forms of antihydrogen such as *antideuterium* (an anti-nucleus containing one antiproton and one antineutron, with a positron

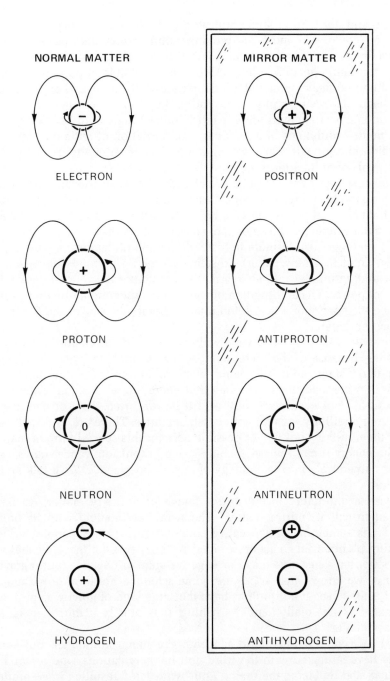

Figure 1-1   An imaginary "magic mirror" shows the difference between normal matter and antimatter (or "mirror matter"). Charge and "handedness" or parity are reversed for positrons and antiprotons. Handedness is reversed for antineutrons. An antihydrogen antiatom would be made of a positron orbiting an antiproton.

in orbit) and *antitritium* (one antiproton and two antineutrons). Following that, one could add more antiprotons, antineutrons, and positrons and make a few atoms of antideuterium, antitritium, and even antihelium-3 (two antiprotons and one antineutron). All of this is possible in principle. In fact, a few hundred heavy nuclei of antideuterium, antitritium, and antihelium-3 have been observed by physicists. Sadly, they have been unable to keep these antimatter fragments under control long enough to add positrons and make neutral antiatoms of antimatter. The formation of antiatoms and the collection and growth of macroscopic amounts of antimatter is one of the major goals of antimatter research today.

The most important characteristic of antimatter is an amazing one: When a matter particle and its mirror matter twin are brought into contact, the two annihilate each other. The mass of both the matter and antimatter is totally converted into energy. The amount of energy that results is given by the most famous mathematical formula in history (and one of the few equations you will ever find in a newspaper), Einstein's $E=mc^2$. This total conversion of matter and antimatter into energy makes antimatter the ultimate potential source of power. The annihilation of a gram of matter and antimatter would produce the energy of a 20-kiloton nuclear bomb, the size of the first ones dropped on Japan.

However, it is unlikely anyone will ever make an antimatter bomb. For one thing, because of the incredible expense of making large amounts of antimatter, a single antimatter bomb would bankrupt the nation that made it. What's more, because of the way that energy is released, an antimatter bomb would not even work very well if it were used. That doesn't mean there are no military uses for antimatter, of course. There are, and we'll talk about them in this book. Other possible uses for this special form of matter, and the energy it can release, include space propulsion, cancer cures, and perhaps an entirely new technology of metallurgy science. We'll talk about those uses, too.

Because antimatter is the mirror image of ordinary matter, we have decided to call it "mirror matter." The word "antimatter," with its prefix "anti-," has something of a negative connotation for some people. Strictly speaking, it's not even an accurate name. Antimatter is not "negative matter." It does not have negative mass, or negative spin, or negative (anti-) gravity (as far as we know—just to be sure, some scientists are even now running experiments to see if antiprotons have the same kind of gravity as protons). Antimatter is not really "anti-" anything. It is merely a mirror image of ordinary matter.

Not *everyone* is uncomfortable with the name antimatter, but some people have struggled with the word and its connotations. One researcher has suggested replacing the prefix "anti-" with "co-," resulting in *co-matter*, *co-protons*, and so forth. Another scientist did some Greek linguistic search-ing, and suggested the prefix "exo-" for "anti-." "Exo" in Greek means "out-side." Other suggestions have included "ob-" (*obmatter*, *obproton*) and "contra-" (*contramatter and contraproton*). None of these suggestions for antimatter have ever caught on with the physics community at large.

Another approach was used by the eminent physicist Hannes Alfven in his 1966 book *Worlds-Antiworlds*. Instead of changing antimatter to another name, Alfven suggested changing the word matter to another term! Since "antimatter"—a coined word—has become a standard term, said Alfven, let's coin a new word for "ordinary" matter. Alfven suggested the word *koinomatter* [pronounced "KOY-no-matter"] after the Greek word *koinos*, meaning common or well-known.

Needless to say, his suggestion has never caught on, either. "Matter" will remain "matter," and "antimatter" will probably remain "antimatter" in the world of physics. However, we feel that "mirror matter" is the most accurate and unbiased term for that form of matter that is the mirror image of "ordinary" matter.

In the Beginning

Mirror matter is, first and foremost, matter. Whether it be positrons, antiprotons, mirror matter hydrogen molecules, or entire mirror galaxies, all mirror matter is still matter. So it's worthwhile taking a look at the history of humanity's increasing understanding of the physical makeup of the cosmos we inhabit.

Today we know that everything in the universe is made of atoms, and that atoms are made of subatomic particles like protons, neutrons, and electrons, and that some of those subatomic particles are in turn composed of still smaller entities called quarks. That, in a nutshell, is the atomic theory of matter. However, this understanding of the construction of the cosmos has not always been ours. Humanity had to discover it, and that meant a lot of thinking, speculating and eventually observing.

The first glimmerings of the atomic theory of matter appeared in the writings of the Greek philosopher Leucippus, who lived around 450 B.C. Leucippus was the founder of the atomist school of philosophy. His most famous student was Democritus, who usually gets the credit for the idea of atoms. Certainly it is the latter's comments on the nature of the world which are the better known. Democritus believed that all of nature was made of tiny, imperceptible but indivisible particles which he and Leucippus called "atoms" (in Greek, *atomos*).

Editor Shmuel Sambursky gives us a fascinating glimpse into the thoughts of Democritus and other pre-scientific philosophers in *Physical Thought from the Presocratics to the Quantum Physicists* (New York: Pica Press, 1975). It's worth reading these comments, for they are the first glimpses of the universe as we understand it today, a vision first seen almost 25 centuries ago: "Democritus [wrote Democritus]" thinks that the nature of the perpetual things consist of small particles infinite in number. . . . He thinks the particles are so small as to be imperceptible to us, and take all kinds of shapes and all kinds of forms and differences of size. Out of them, like out of elements [earth, air, fire, water], he now lets combine and originate the visible and perceptible bodies. They move in confusion in the void because of their dissimilarity and the other specified differences, and while in motion they collide and become interlocked in entanglement of a kind which causes them to be juxtaposed and in proximity to one another without actually forming any real unity whatsoever. The cause of the continuance of aggregations of particles for some period of time, he says, is their fitting into one another in bondage—for some of them are uneven, others barbed, some concave and some convex and others possessed of innumerable other distinctive qualities. He thinks they hold together and continue to do so until the time when some stronger force coming from the environment disrupts and disperses them in different fashion.

Democritus also believed that true knowledge of the cosmos could come only through thought. The evidence of the senses was often faulty, he wrote, and so could not be trusted. That attitude was a common one in the ancient Greek culture. It may have been responsible for the fact that few if any of the great Greek thinkers ever tried to test their philosophical proposals with real-life experiments. If they had—if the Greek culture of the time had been accepting of the evidence of the senses gained from experiments—world history might well have turned out quite differently.

Democritus's ideas of the atom were echoed and at times amplified by other Greek philosophers, including Plato and Epicurus. Plato, for example, proposed a doctrine of geometrical atomism, postulating the essential particles of the cosmos to have regular geometrical shapes: the pyramid (made of four equilateral triangles), the octahedron (eight equilateral triangles), the icosahedron (120 triangles), and the cube (six quadrangles). He proposed that each of these four forms represented one of the four elements: fire (the pyramid, which was sharp and cutting), air (octahedron); water (icosahedron); and earth (cube). "Now," Plato wrote,

we must think of all these bodies as so small that a single body of any one of these kinds is invisible to us because of its smallness; though when a number are aggregated the masses of them can be seen. And with regard to their numbers, their motions, and their powers in general, we must suppose that the god adjusted them in due proportion, when he had brought them in every detail to the most exact perfection permitted by Necessity willingly complying with persuasion.

Plato lived from about 428 to 320 B.C., partly contemporaneous with Democritus. Epicurus was born about 20 years before Plato died, about the year 341 B.C., and died about 271 B.C. He founded the Epicurean school of philosophy, which adopted and amplified Democritus's atomistic philosophy.

Some of Epicurus's statements about atoms and the universe sound remarkably modern:

> Of bodies, some are composite, others the elements of which these composite bodies are made. The elements are indivisible and unchangeable, and necessarily so, if things are not all to be destroyed and pass into non-existence, but are strong enough to endure when the composite bodies are broken up, because they possess a solid nature and are incapable of being anywhere or anyhow dissolved. It follows that the first beginnings must be indivisible, corporeal bodies. ... the sum of things is unlimited both by reason of the multitude of the atoms and the extent of the void.... The atoms are in continual motion through all eternity. Some of them rebound to a considerable distance from each other, while others merely oscillate in one place when they chance to have got entangled or to be enclosed by a mass of other atoms shaped for entangling.

The Roman poet Titus Lucretius Carus lived more than two hundred years after Epicurus died, but his epic poem De Rerum Natura (On the Nature of Things) is one of the best sources of information on the Epicurean view of nature. Lucretius wrote extensively of atoms in De Rerum Natura. At one point he spoke of the visible effects of invisible entities:

> Garments hung up on a surf-beaten shore grow damp, the same spread in the sun grow dry; yet none has seen either how the damp of the water pervaded them, or again how it departed in the heat. Therefore the water is dispersed into small particles, which the eye cannot in any way see. Moreover, with many revolutions of the Sun's year, a ring on the finger is thinned underneath by wear, the fall of drippings hollows a stone, the curved ploughshare of iron imperceptibly dwindles away in the fields, and the stony pavement of the roads we see already to be rubbed away by men's feet; again bronze statues set by the gateways display the right hands thinned away by the often touching of those who kiss and those who pass by. These therefore we observe to be growing less because they are rubbed away; but what particles are separated on each occasion, our niggardly faculty of sight has debarred us from seeing. Lastly, whatever time and nature little by little adds to things, compelling them to grow greater by degrees, no keenness of sight, however strained, can perceive; nor further when things grow old by age and wasting, nor when rocks hanging over the sea are eaten away by the fine salt, could you discern what they lose upon each occasion. Therefore nature works by means of bodies unseen.

Lucretius's words are a vivid evocation of sensory observation leading inescapably to the conclusion that "nature works by means of bodies unseen." How tantalizingly close the great poet came to the core of true science, observation plus experimentation! Yet neither the Greeks nor the

Romans were quite able to make that leap. And so the atomistic theory of nature slipped into obscurity. The philosophical meanderings of Democritus and Epicurus and the poetic observations of Lucretius came to naught.

The Modern Era Begins

The modern atomic theory of matter truly began at the start of the nineteenth century, with the work of the English scientist John Dalton. Dalton revived Democritus's concept of indivisible atoms as the fundamental particles of matter. He applied the idea in a practical fashion, using it to construct a table or arrangement of *atomic weights*. He also used the concept of atoms and atomic weight to devise an explanation of gas pressure now called Dalton's Law. The total pressure of a mixture of different gases, said Dalton, is equal to the sum of the pressures of the gases in the mixture, each acting independently of the others. Dalton's Law is, in fact, correct, and is only explainable if the different gases are made of different and discrete particles—atoms. The different atoms of different gases, of course, have different weights, and so end up exerting different pressures.

Dalton's pioneering writings on atoms were published in 1808, in the book *A New System of Chemical Philosophy*. Dalton based his ideas of atomic composition of matter on observation, experiment, logical thinking, and empirical evidence. "A very familiar instance is exhibited to us by water," Dalton wrote,

> a body, which, in certain circumstances, is capable of assuming all the three states. In steam we recognize a perfectly elastic fluid, in water, a perfect liquid, and in ice a complete solid. These observations have tacitly led to the conclusion which seems universally adopted, that all bodies of sensible magnitude, whether liquid or solid, are constituted of a vast number of extremely small particles, or atoms of matter bound together by a force of attraction, which is more or less powerful according to circumstances. . . . Whether the ultimate particles of a body, such as water, are all alike, that is, of the same figure, weight, etc., is a question of some importance. From what is known, we have no reason to apprehend a diversity in these particulars: If it does exist in water, it must equally exist in the elements constituting water, namely, hydrogen and oxygen. Now it is scarcely possible to conceive how the aggregates of dissimilar particles should be so uniformly the same. If some of the particles of water were heavier than others, if a parcel of the liquid on any occasion were constituted principally of these heavier particles, it must be supposed to affect the specific gravity of the mass, a circumstance not known. Similar observations may be made on other substances. Therefore we may conclude that the ultimate particles of all homogeneous bodies are perfectly alike in weight, figure, etc. In other words, every particle of water is like every other particle of water; every particle of hydrogen is like every other particle of hydrogen, etc.

Today physicists and chemists define atoms somewhat differently than did Democritus, and more as Dalton did, in chemical terms: An atom is the smallest particle having the unique chemical characteristics of an element. An element, in turn, is a substance which cannot be broken down into a simpler substance by chemical reactions. About 90 elements exist in nature; the total number of known elements, including those made by humans in recent years, is 106 (Table 1–4). They range from hydrogen, the simplest element known; to uranium, the heaviest element known in nature; to element 106, the heaviest known element of all. The most abundant element on Earth is oxygen, followed closely by carbon and silicon. In the universe as a whole, though, hydrogen is the most abundant element. More than 80 percent of all the matter in the universe is hydrogen. Most of the rest is helium, the second lightest element and the second most abundant in the cosmos. In 1869 the Russian physicist Dmitri Mendeleev discovered a relationship among the physical and chemical characteristics of elements when they were arranged by atomic number. The atomic number of an element is the number of protons in its nucleus; thus hydrogen has an atomic number of 1, oxygen of 8, and uranium of 92. The periodic table (Table 1–5) is a graphic representation of such an arrangement. Mendeleev himself realized that his table had some "holes." There were undiscovered elements lurking somewhere. He predicted the existence of one, which he dubbed "eka-boron," and suggested its characteristics. "Eka-boron" was discovered in 1879 by Lars Nilson and named scandium. It has an atomic number of 21 and lies between calcium and titanium on the periodic table. Mendeleev lived to see his prediction confirmed. Many other elements have been found since the conception of the periodic table, and they all fit in.

By John Dalton's time scientists knew that some substances were made of still simpler constituents. Water was already known to be composed of hydrogen and oxygen, two atoms of hydrogen combined with one of oxygen. Water was a *compound,* a *molecule.* Hydrogen and oxygen were *elements*—in Dalton's view and the view of other scientists, each made of a single type of *atom,* the ultimate, indivisible particle of matter that Democritus had spoken of more than 2,400 years earlier. Dalton himself thought of atoms as featureless spheres, while some of his followers imagined them has having little hooks, so as to explain the chemical combinations of atoms into various molecules. However, the idea that the atom is the ultimate particle was just too simple to be true, and observation and experiment showed that it could not be the case.

The phenomenon of electricity has been known to humanity in one form, static electricity, for thousands of years. The American scientist Benjamin Franklin had experimented with it more than 50 years before Dalton's work, as had Englishman Stephen Grey in the 1730s. Franklin explained electricity as a flow of a "negative fluid."

It was not long before people were constructing "Leyden jars," crude forms of capacitors in which to store the electrical "fluid." Other people developed batteries, essentially stacks of dissimilar metals that generated

Table 1-4

A list of the 106 known elements, arranged by atomic number from the lightest (hydrogen) to the heaviest (element 106).

Atomic Number	Element	Symbol	Atomic Mass (u)
1	hydrogen	H	1.00797
2	helium	He	4.0026
3	lithium	Li	6.939
4	beryllium	Be	9.0122
5	boron	B	10.811
6	carbon	C	12.01115
7	nitrogen	N	14.0067
8	oxygen	O	15.9994
9	fluorine	F	18.9984
10	neon	Ne	20.183
11	sodium	Na	22.9898
12	magnesium	Mg	24.312
13	aluminum	Al	26.9815
14	silicon	Si	28.086
15	phosphorus	P	30.9738
16	sulfur	S	32.064
17	chlorine	Cl	35.453
18	argon	Ar	39.948
19	potassium	K	39.102
20	calcium	Ca	40.08
21	scandium	Sc	44.956
22	titanium	Ti	47.90
23	vanadium	V	50.942
24	chromium	Cr	51.996
25	manganese	Mn	54.9380
26	iron	Fe	55.847
27	cobalt	Co	58.9332
28	nickel	Ni	58.71
29	copper	Cu	63.54
30	zinc	Zn	65.37
31	gallium	Ga	69.72
32	germanium	Ge	72.59
33	arsenic	As	74.9216
34	selenium	Se	78.96
35	bromine	Br	79.909
36	krypton	Kr	83.80
37	rubidium	Rb	85.47
38	strontium	Sr	87.62
39	yttrium	Y	88.905
40	zirconium	Zr	91.22
41	niobium	Nb	92.906
42	molybdenum	Mo	95.94
43	technetium	Tc	[99]
44	ruthenium	Ru	101.07
45	rhodium	Rh	102.905
46	palladium	Pd	106.4
47	silver	Ag	107.870
48	cadmium	Cd	112.40
49	indium	In	114.82
50	tin	Sn	118.69
51	antimony	Sb	121.75
52	tellurium	Te	127.60
53	iodine	I	126.9044

Atomic Number	Element	Symbol	Atomic Mass (u)
54	xenon	Xe	131.30
55	cesium	Cs	132.905
56	barium	Ba	137.34
57	lanthanum	La	138.91
58	cerium	Ce	140.12
59	praseodymium	Pr	140.907
60	neodymium	Nd	144.24
61	promethium	Pm	[145]
62	samarium	Sm	150.35
63	europium	Eu	151.96
64	gadolinium	Gd	157.25
65	terbium	Tb	158.924
66	dysprosium	Dy	162.50
67	holmium	Ho	164.930
68	erbium	Er	167.26
69	thulium	Tm	168.934
70	ytterbium	Yb	173.04
71	lutetium	Lu	174.97
72	hafnium	Hf	178.49
73	tantalum	Ta	180.948
74	tungsten	W	183.85
75	rhenium	Re	186.2
76	osmium	Os	190.2
77	iridium	Ir	192.2
78	platinum	Pt	195.09
79	gold	Au	196.967
80	mercury	Hg	200.59
81	thallium	Tl	204.37
82	lead	Pb	207.19
83	bismuth	Bi	208.980
84	polonium	Po	[210]
85	astatine	At	[210]
86	radon	Rn	[222]
87	francium	Fr	[223]
88	radium	Ra	226.05
89	actinium	Ac	[227]
90	thorium	Th	232.038
91	protactinium	Pa	[231]
92	uranium	U	[238.03]
93	neptunium	Np	[237]
94	plutonium	Pu	[242]
95	americium	Am	[243]
96	curium	Cm	[247]
97	berkelium	Bk	[247]
98	californium	Cf	[249]
99	einsteinium	Es	[254]
100	fermium	Fm	[257]
101	mendelevium	Md	[256]
102	nobelium	No	[259]
103	lawrencium	Lw	[260]
104	(unnamed)		[261]
105	(unnamed)		[262]
106	(unnamed)		[263]

Table 1-5

The Periodic Table of Elements, devised orginally by Mendeleev, depicts the different relationships among the elements.

Group	I	II					Transition Elements						III	IV	V	VI	VII	O
Period 1	1 H 1.00797																	2 He 4.0026
2	3 Li 6.939	4 Be 9.0122											5 B 10.811	6 C 12.01115	7 N 14.0067	8 O 15.9994	9 F 18.9984	10 Ne 20.183
3	11 Na 22.9898	12 Mg 24.312											13 Al 26.9815	14 Si 28.086	15 P 30.9738	16 S 32.064	17 Cl 35.453	18 Ar 39.948
4	19 K 39.102	20 Ca 40.08	21 Sc 44.956	22 Ti 47.90	23 V 50.942	24 Cr 51.996	25 Mn 54.9380	26 Fe 55.847	27 Co 58.9332	28 Ni 58.71	29 Cu 63.54	30 Zn 65.37	31 Ga 69.72	32 Ge 72.59	33 As 74.9216	34 Se 78.96	35 Br 79.909	36 Kr 83.80
5	37 Rb 85.47	38 Sr 87.62	39 Y 88.905	40 Zr 91.22	41 Nb 92.906	42 Mo 95.94	43 Tc (99)	44 Ru 101.07	45 Rh 102.905	46 Pd 106.4	47 Ag 107.870	48 Cd 112.40	49 In 114.82	50 Sn 118.69	51 Sb 121.75	52 Te 127.60	53 I 126.9044	54 Xe 131.30
6	55 Cs 132.905	56 Ba 137.34	57–71 *	72 Hf 178.49	73 Ta 180.948	74 W 183.85	75 Re 186.2	76 Os 190.2	77 Ir 192.2	78 Pt 195.09	79 Au 196.967	80 Hg 200.59	81 Tl 204.37	82 Pb 207.19	83 Bi 208.980	84 Po (210)	85 At (210)	86 Rn (222)
7	87 Fr (223)	88 Ra (227)	(89–103) †	(104)	(105)	(106)												

*Lanthanide rare–earth elements

57 La 138.91	58 Ce 140.12	59 Pr 140.907	60 Nd 144.24	61 Pm (145)	62 Sm 150.35	63 Eu 151.96	64 Gd 157.25	65 Tb 158.924	66 Dy 162.50	67 Ho 164.930	68 Er 167.26	69 Tm 168.934	70 Yb 173.04	71 Lu 174.97

† Actinide rare–earth elements

89 Ac (227)	90 Th 232.038	91 Pa (231)	92 U 238.03	93 Np (237)	94 Pu (242)	95 Am (243)	96 Cm (245)	97 Bk (249)	98 Cf (249)	99 Es (254)	100 Fm (252)	101 Md (256)	102 No (254)	103 Lw (257)

electricity. They were the forerunners of the batteries that today power everything from toy rayguns to orbiting satellites. One such battery used plates of copper and zinc dipped in a jar of sulphuric acid. The copper and zinc atoms—with no electrical charge—were being put together in a way that produced an electrical charge. The electrical charge (or fluid, if you were Ben Franklin) had to come from somewhere. The only place it could possibly come from was from somewhere *in the atoms themselves.* Somewhere in the zinc or copper atoms was something with a negative electrical charge. Atoms are normally neutral in charge. Therefore, there must also be something inside them with a *positive* electrical charge, so that the charges all balanced out. Atoms were not indivisible. They must have structure. They must have smaller charged constituents within them.

The first such "subatomic" particle to be discovered was the electron, the negatively charged particle responsible for Franklin's "negative fluid." English scientist J.J. Thomson discovered and named the electron in 1897, some 89 years after Dalton propounded his theory of atoms as the ultimate particles. The shape of modern atomic theory was beginning to unfold, and that unfolding would come more and more quickly in the twentieth century.

Chapter Two

From Matter to Its Mirror Image

The first two decades of the twentieth century saw an explosion of knowledge in physics. The goal of physics is to determine what the universe is composed of and what its governing laws consist of. The discovery of mirror matter was therefore a natural outgrowth of the physicists' search for an understanding of nature. Its discovery arose directly from the two greatest developments of twentieth-century physics, *relativity* and *quantum mechanics*. *Special relativity*, formulated in 1905 by Albert Einstein, explained the relationship of space to time, and of space-time to the speed of light, as well as how objects act when they travel at extremely high velocities. Three of Einstein's findings are important to the discovery and use of mirror matter. First, as a material object in motion approaches the speed of light, that object's mass begins to increase. Second, for an observer located outside that object (in another frame of reference, as Einstein would say) the passage of time for that object appears to slow down. Third, the relationship between mass, energy, and the speed of light could be expressed by the now-famous mathematical equation, $E=mc^2$.

By the mid-1920s Einstein's Special Theory of Relativity had become accepted as accurate by most physicists. Some of its implications seemed a bit bizarre, but even the general public tended to accept relativity. "Everything's relative" became a commonly used phrase. It was something that made a certain amount of sense.

This was not so true of quantum mechanics. Scientists at the turn of the century had been faced with a serious quandary. Experiments probing the deep nature of matter and energy had uncovered discrepancies between theory and reality. For example, different experiments showed that light not only acted like it was made of waves; sometimes it appeared to be made of tiny particles. This didn't make sense. Almost in desperation, physicist Max Planck came up with an explanation. Light waves, said Planck in 1901, must come in little packets, which he named *quanta*. Light was neither wave-like nor particle-like, but instead was made of something that was both wave-like and particle-like.

There was nothing commonsensical about this at all. However, by the 1920s physicists were trying to apply this developing theory not only to energy like light, but also to the atom and its particles. If light waves sometimes acted like particles, then perhaps some subatomic particles might also act like waves. In particular, physicists were interested in the wave-like nature of the electron. In 1926 physicist Erwin Schrödinger presented the world with a new and powerful mathematical vision of subatomic particles. Schrödinger's theory first dealt mathematically with subatomic entities like electrons as if they were waves. Then, the results of the math were interpreted in such a way as to predict the *probability* of the entity being present as a particle at a particular point in space.

What does it mean to describe the motion of a particle as a probability wave? It means you never know for sure where it is going to be, or what state it will be in when it gets there. One real-life example of the latter is the notion of *half-life*. Physicists talk of radioactive atoms as having a half-life of some amount of time. Half-life is the average amount of time required for one-half of any quantity of identical radioactive atoms to undergo radioactive decay. The radioactive atom cobalt-60 has a half-life of 5.3 years; of one gram of cobalt-60, one-half gram of it will have undergone radioactive decay in 5.3 years. When another 5.3 years have passed, one-half of the remaining half-gram—or one-quarter gram—will have undergone radioactive decay . . . and so on. You cannot predict when a *particular atom* of cobalt-60 will decay. You *can* predict, with great accuracy, how many atoms *in a large sample* will do so in a particular time period. Predicting half-lives is no problem. Predicting when a specific atom will decay radioactively is impossible.

With the publication of Schrödinger's mathematics, all the physicists in the world had to become statisticians. Some didn't like this way of dealing with reality, including Einstein. He is supposed to have once said, "God does not play dice with the universe." However, Schrödinger's theory worked, and worked well. God does indeed play dice with the cosmos. This turned philosophy upside down. The simple mechanical universe, where everything was in its place and there was a place for everything, was gone. Instead, scientists now envisioned a universe in which anything was possible. Predestination was out. Free will was in. This logic was perhaps glimpsed by Lucretius more than 2,000 years earlier, when he wrote that "the fact that the mind itself has no internal necessity to determine its every act and suffer in helpless passivity—this is due to the slight swerve of the atoms at no determinate time or place."

Schrödinger's equations were quite powerful and accurately predicted what was observed in various experiments carried out on atoms and sub-atomic particles like protons and electrons. Along with Werner Heisenberg, who was working on the same question, Schrödinger had succeeded in bringing into existence a fully developed quantum theory of physics. However, Schrödinger's theory was not *relativistic*. It only applied to systems of particles like electrons that were moving at *low velocities* and not velocities close to the speed of light. It also did not take into account the electron's spin.

A young physicist named Paul Dirac set out to remedy these shortcomings. He found a way to combine the Schrödinger equations for quantum mechanics, the Einstein equations for special relativity, the Maxwell equations for electromagnetism, and his own non-relativistic equations for the behavior of the electron into a single set of equations. This new set of equations described the relativistic quantum behavior of the spinning electron. In December 1929, Dirac's solution was published in the *Proceedings of the Royal Society*. It was a startling paper, for Dirac's equations did something rather bizarre. In the classical physics of Newton, the energy of a particle always has a positive value. The equations never give the result of "negative energy" for a particle. Dirac's new equations for the electron, however, had two possible solutions: an electron with positive energy, *or an electron with negative energy.*

This seemed absurd, but Dirac could not simply throw out his work and try again. His equations happened to explain perfectly some odd experimental observations of atomic and subatomic behavior that had been bedeviling physicists. Dirac had to admit there were indeed two solutions to the equations, and that one of them implied the existence of electrons with negative energy. If that was so, perhaps it would be interesting to try explaining how an electron with negative energy would act. Dirac discovered that an electron with *negative* energy passing through a magnetic field would act exactly like an electron with *positive* energy—if the electron had a positive instead of negative electric charge.

To Dirac, this implied that for every particle that existed there was a corresponding mirror-image antiparticle. Dirac first suggested that the electron's antiparticle was the proton. J.J. Thomson, who had discovered the electron in 1897, actually observed protons in a series of experiments he started in 1906. He identified them as positively charged hydrogen atoms—which in fact *are* protons. At the time, though, Thomson didn't realize that they were discrete subatomic particles, like electrons. He knew atoms were made of positively and negatively charged parts. However, he envisioned the atom as a "plum pudding," with the electrons being negatively charged plums embedded in a diffuse positively charged pudding. Thomson's theory could be tested by bombarding materials with charged particles and seeing how those charged particles were deflected. Those experiments were carried out by Ernest Rutherford, one of the most prominent researchers in the field of radioactivity. One of Rutherford's earlier discoveries had been the emission by some heavy radioactive elements of *alpha particles.* Today physicists know that alpha particles are actually the nuclei of helium atoms. During Rutherford's day they were only known to be particles with a positive charge twice as great as (and opposite to) an electron, and about 7,000 times as heavy as an electron. Rutherford decided to bombard thin films of gold with alpha particles and observe how the alpha particles were scattered as they hit the atoms in the foils. He began these experiments in 1907, a year before he received the Nobel Prize in chemistry for his research into radioactivity.

If Thomson's atomic model were correct, the alpha particles would be barely disturbed in their passage through the metallic film targets. The electrons in the gold atoms would deflect a few alpha particles, but only slightly. Rutherford and his colleagues discovered that most of the alpha particles were indeed scattered through small angles. But not all. A significant number of alpha particles were deflected through large angles. Some alpha particles actually ricocheted right out of the gold foil and back toward the alpha particle source. Rutherford later said it was like firing a 15-inch shell at a piece of tissue paper and having it bounce back. Thomson's model could not be correct. It seemed that most of an atom consisted of a tiny, positively charged nucleus. The electrons must somehow surround this nucleus—perhaps in a kind of orbit—at a very great distance. Thus was born the nuclear model of the atom, which is essentially the picture of the atom still accepted today.

In 1919, Rutherford finally made a positive identification of protons. As Thomson and others had earlier deduced, the proton has a positive electrical charge with exactly the opposite strength of the electron's negative charge. For that reason, it seemed sensible for Paul Dirac to first call the electron's antiparticle a proton. However, it wasn't that easy. The proton weighs more than 1,836 times the electron. How could the one be the mirror image of the other? In a second paper published in 1931, Dirac admitted that the proton really couldn't be the electron's antiparticle. Robert Oppenheimer and Hermann Weyl had already suggested that the negative energy electron must be a new type of particle. Dirac agreed, and proposed that it be called the *antielectron*. Later it was named the *positron*. Dirac noted that it would not be easy to detect the antielectron (or positron), since it would have a strong tendency to recombine with any nearby normal electron. In the same 1931 paper, Dirac also predicted that the proton would have a mirror particle.

Electron Holes and Time Travel

Mirror matter particles were not the only explanation Dirac suggested for the positive and negative energy results of his equations. Dirac suggested another model, in which the positron is a "hole" in the infinite sea of electrons with negative energy. We don't see any of the negative-energy electrons Dirac's theory predicted because all of the possible negative-energy states are filled with negative-energy electrons. Since all the states are full, there are no differences to be observed, so we cannot detect this infinite sea of negative-energy electrons.

Here's one way to picture this strange theory. Imagine that your uncle in Seattle has sent you a box of genuine Washington apples. Twenty-five apples fit snugly on the bottom of the box, and another five sit on top of the bottom layer. Now imagine that the 25 apples on the bottom represent electrons with negative energy. The other five apples, the ones sitting on top of the first 25, represent electrons with positive energy—that is, regular electrons that we encounter every day.

Normally you'd take one of the five apples from the upper layer, since they're easy to get at. The 25 apples in the bottom of the box are packed so tightly that you'd have to exert quite a bit of energy to pry one out. Perhaps you'll just ignore them, then.

But no. You are curious about those negative-energy apples in the bottom layer. So you take the time and effort to pull one out. You extract a Red Delicious from the bottom layer of 25 apples. The apples in the bottom row are packed so tightly that they don't shift places as you remove the one apple. This is akin to removing a negative-energy electron. In its place is a hole. If the apple represents an electron with negative energy, then logically the *hole* you have created has *positive* energy. In fact, you can think of the hole as the equivalent of an anti-apple (or antielectron) with positive energy, since it corresponds to the *absence* of an apple (electron) with *negative* energy. So the hole is essentially the same as a positive-energy anti-apple.

In the process of pulling out the apple from the bottom layer, however, you have jiggled those five loose apples in the upper layer. These are the apples that represent normal, positive-energy electrons. You jiggle them so much, in fact, that one of them rolls into the hole left behind when you pulled out the negative-energy apple. So the hole is now gone. *But so is the apple.* It has filled the hole and is now firmly embedded in the bottom layer of "negative-energy" apples.

That's what would happen, said Dirac in 1931, if you created an antielectron or "hole." A normal electron would quickly encounter the antielectron hole. The electron would fall into the hole to fill it, and both the electron and the hole would disappear from view. Energy would be released when this happened. The amount of energy would be equal to the difference in energy between a positive-energy electron and a negative-energy electron. That in turn is the same as two electron masses converted entirely into energy.

The antiparticle theory and the electron hole theory are just different ways of looking at the same thing. Using this "hole" model, Dirac was not only able to account for the negative energy states in the solutions to his equations that predicted the existence of the antielectron, but he was able to predict that the annihilation process would take place.

There's still another way of looking at a positron, suggested by Nobel Prize-winning physicist Richard Feynman. A positron, Feynman has written, can also be thought of as an electron *moving backwards in time!* An electron doing such a remarkable thing would be indistinguishable from an electron with a positive charge. This would also be essentially true of any other mirror matter particle or object. Of course, this doesn't prove the existence of time travel, nor does it mean that mirror matter particles are time travelers. It just means that they could be. The Feynman pictorial model (and the mathematical equations for that model) is still another way of explaining the observed actions of electrons and positrons.

Perhaps, though, time-traveling electrons aren't strange enough for you. An even more bizarre extension of the Feynman model has been suggested by John Wheeler. It goes like this: (1) All electrons are produced initially by particle creation of an electron-positron pair. That is, you can't make an electron without also making a positron at the same time. (2) Also, all electrons and positrons ultimately disappear in a positron-electron annihilation process. (3) So all the paths that electrons and positrons travel through space-time begin and end with the beginning or end of some other electron or positron. (4) That means all the paths are linked end-to-end into one long trajectory that zigzags back and forth in space and time. (5) Therefore: if the positron is just an electron moving backwards in time, then all the electrons and positrons we now observe in the universe *are just one single electron seen at different portions of a single long electron path!*

This truly strange model of the electron and positron also explains why all electrons have exactly the same charge. It's the same electron!

The Antielectron Game

Dirac's antimatter explanation of his relativistic quantum equations created quite a stir, especially among physicists studying *cosmic rays*. Cosmic rays had been discovered in 1911 by Viktor Hess. They were the only form of energy known in the 1920s and 1930s powerful enough to create an antielectron. The theoretical antielectron would be identical to the electron in every way but electrical charge. Its mass would be the same. The problem was that physicists possessed no method or machine in the late 1920s with that much energy. Nature, however, did. Cosmic rays are the nuclei of atoms traveling at nearly the speed of light and possessing enormous energies. Most cosmic rays are actually the nuclei of hydrogen atoms—that is, protons. These primary cosmic rays smash into the upper atmosphere from outer space continuously, breaking atoms apart into subatomic fragments and releasing gamma rays and X rays. These in turn hit and break apart atoms and molecules further down in the atmosphere. The eventual result is a shower of secondary cosmic rays and their debris. The cascade of other high energy particles eventually reaches the Earth's surface and can be detected and measured.

By 1929 physicists knew that some cosmic rays had energies as high as 15 million *electron volts*. The electron volt (or eV) is the unit of energy used by particle physicists. If a metal plate has a positive voltage of one volt (a regular flashlight battery produces 1.5 volts), then an electron will be attracted to that plate. Just before the electron reaches the plate, it will have an energy of one electron volt (1 eV). Some scientists reasoned that such highly energetic cosmic rays could be producing the still-undiscovered antielectrons Dirac predicted in 1929. For example, if a cosmic ray with over a million electron volts of energy hit an atom, it might well create a particle pair—an electron and an antielectron. Detecting such "pair production" events would prove the existence of antimatter particles.

In 1932 the United States was in the throes of the Great Depression. Distraction from misery came from movies—and from baseball, which had become the Great American Pastime. Babe Ruth was King. Like the motion of Lucretius's atoms, one could never be sure when the Babe's bat would send a ball into the stands, but statistically it happened about once every three games. Ruth played for the New York Yankees at the time, and they were about to become the best team in baseball. (When Ruth left the Yankees's lineup, the "hole" was worth about one run per game for the other teams in the American League—a vivid example of how something negative can have concrete consequences.) The Bronx Bombers had played in four of the last ten World Series, and won three of them. In 1932 the Yankees swept the Chicago Cubs four games to zip in the October classic.

That was in New York City. Three thousand miles to the west, in the rural Southern California town of Pasadena, a tiny private college named the California Institute of Technology (Caltech) had taken something else from Chicago. Actually, it was a *someone* else: Robert Millikan, the Nobel Prize-winning physicist (Figure 2-1). Millikan's job was to turn Caltech's tiny

Figure 2-1   Robert Millikan.
(Courtesy California Institute of Technology Archives)

physics department into a big one, one that made big discoveries and won big prizes. His efforts soon produced spectacular results. One of Millikan's colleagues was a 27-year-young professor named Carl Anderson. In 1932 Millikan and Anderson were investigating cosmic rays, and they had built a large *cloud chamber*—an early kind of particle detector (Figure 2–2). When subatomic particles passed through the chamber, they left ghostly vapor trails in the supersaturated air within. The two researchers placed powerful magnets around it to blanket the interior of the chamber with a magnetic field. Like others, they saw that the cosmic rays detected in the chamber were bent by the field. Obviously they were made of charged particles, and the direction and thickness of the paths seen in the cloud chamber revealed the mass of the particles—and their charge. Anderson took some photos of the paths seen in the cloud chamber, and examined them closely (Figure 2–3). He noticed that some of the trails were exactly like those made by electrons, but were curved by the magnetic field in the opposite direction. At this point Anderson was not aware of Dirac's prediction of the existence of an electron with negative energy. The tracks were quite baffling. But he finally came to the tentative conclusion that they were created by cosmic

Figure 2-2 Carl Anderson sits next to the cloud chamber in which he detected the first positron. (*Courtesy California Institute of Technology Archives*)

ray particles that were positively charged, and with the mass of an electron. Anderson continued his work with the cloud chamber, examining photos of the vapor tracks left behind by cosmic rays. After nearly a year of effort and observation, he firmly identified instances of the pair-production of electrons and antielectrons from the impact of cosmic rays. Even when he and others finally connected the photos with Dirac's prediction, acceptance of the existence of antimatter still came slowly for some physicists. The evidence, however, was clear. Not only did the photos prove the existence of antimatter; they also proved the validity of Einstein's formula $E=mc^2$ by providing direct evidence of the conversion of energy into mass!

The same year that Anderson spotted the positron trails in the cloud chamber photos (and the Yankees swept the Cubs), British scientist Sir James Chadwick, a colleague of Rutherford's, discovered the neutron. Like Rutherford's proton, the neutron was a *nucleon*, a subatomic particle that resided in the nucleus of atoms. It had just slightly more mass than a proton, and no electrical charge.

The following year the Yankees finished out of the running in the American League. In the physics game, though, Paul Dirac and Erwin Schrödinger won the Nobel Prize for their pioneering work in developing quantum physics. Three years later, in 1936, Carl Anderson received the Nobel Prize for his discovery of the electron's mirror matter particle, which he had named the positron (Figure 2–4). He shared the Nobel money with Viktor Hess, the discoverer of the cosmic rays that created the positrons Anderson had discovered. It was an appropriate "pair-production" of the preeminent scientific award. Caltech's physics department had its first "in-house" Nobel Prize. That year the Yankees were back in the World Series, beating the National League's New York Giants four games to two in the first of four straight winning appearances in the Series.

Paul Dirac went on to make many other important contributions to physics, including his still-controversial prediction of the existence of *magnetic monopoles*. A magnetic monopole would be a particle that carried either a north or south magnetic pole—but not both at once. Particles with a positive or negative *electric charge* are commonplace, and a magnetic monopole would be analogous to them. So far, though, no definite physical proof exists for monopoles. Still, Dirac is probably most famous for his prediction of the existence of antimatter.

Among those who knew him, he was also known as an extremely shy person. There are many stories that illustrate his reticence. Heinz Pagels relates one in his book *Perfect Symmetry*. It seems Dirac was once at a dinner, and was seated next to Richard Feynman, a Nobel Prize winner and one of the giant figures in modern physics for his development of the theory of quantum electrodynamics—also an inveterate talker, and anything but shy. Apparently Feynman turned to Dirac and asked him, "Did you feel good when you wrote down that equation?"—the one that led to the discovery of antimatter. At first Dirac said nothing. Finally he replied, "Yes." Another very long pause, and then Dirac said, "Are *you* working on an

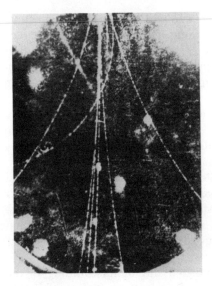

Figure 2-3   An early photograph by Anderson of electron and positron tracks in the cloud chamber. The shower of particles comes down from above. The electrons are curving to the left, the positrons to the right, bent by a powerful magnetic field. *(Courtesy California Institute of Technology Archives)*

Figure 2-4   Carl Anderson, winner of the Nobel Prize for discovering the first particle of mirror matter, the positron. *(Courtesy California Institute of Technology Archives)*

equation, too?" So startled was Feynman by this utterly surprising response that, for once, he had nothing to say.

The positron's discovery was exciting and tantalizing—exciting, because it confirmed Dirac's work, revealed the validity of quantum physics, and showed how astronomy and physics would eventually become inextricably entwined; tantalizing, because if mirror matter electrons existed, then surely there must be antiparticles to the proton as well. Perhaps even the neutron had an antiparticle.

Discovery of further mirror matter particles, however, would have to wait for a while—a long while. The energy-mass of the proton and neutron is equal to about 938 million electron volts. Not even the newfangled cyclotron "atom smasher," invented by Ernest O. Lawrence in 1930, could smash atoms with that much energy. The next mirror matter discovery would not come for another 22 years.

The Big Science Game

In 1954, driven by the pressures of World War II, the Manhattan Project, and America's nearly unchallenged world preeminence, the world of science witnessed the development of "Big Physics." At the University of California at Berkeley, physics was Very Big. On a eucalyptus-covered hill above the campus, the legendary Ernest Orlando Lawrence had assembled the world's most illustrious collection of physicists.

The idea of searching for antiprotons was in the air in the early months of 1954. At meetings and conferences, in classrooms and laboratories, physicists were discussing the possibility of creating and detecting the proton's mirror particle. It had been 25 years since Paul Dirac had suggested the existence of positive electrons, 22 years since Anderson and Millikan had found the positron in their cloud chamber photos. It seemed obvious that if antielectrons existed, then antiprotons should exist, too. Yet no one had found them. The reason was that the energy needed to create antiprotons was greater than any existing particle accelerator could reach. By 1954, though, that "energy wall" had been breached. What was needed now were specific schemes to find the antiproton, the proper detectors, and some running time on an atom smasher powerful enough to do the trick.

Time, energy, and genius began to come together in several places around the world. One of the teams making serious plans was in Berkeley. It was headed by the Italian physicist Emilio Segrè, 50 years old and already famous for his work in physics. Also on the team were three Americans: Clyde Wiegand, a member of the laboratory staff and a genius at building experimental equipment; Thomas Ypsilantis, a graduate student; and 35-year-old physicist Owen Chamberlain. Tall and skinny, Chamberlain looked like he should have been on the college basketball team rather than tinkering with the Bevatron on the hill above the university. The Bevatron was the powerful new particle accelerator that Lawrence and his associates had put together to probe the innermost secrets of the atom (Figure 2–5). At the beginning of 1954 the giant doughnut of vacuum pipes, lead shielding, and massive magnets was one of the world's most powerful atom smashers (Figure 2–6).

We mentioned that the electron volt, or eV, is the unit of energy used by physicists. A distant electron attracted to a metal plate with a positive charge of one volt will have a kinetic energy (that is, energy of motion) of one electron volt just before it hits the plate. A proton has a positive charge, so it would be attracted to a plate with a negative voltage of minus one volt. The charge on the positively charged proton is the same magnitude as the charge on the negatively charged electron, so the proton would also gain an energy of 1 eV. But the proton is heavier than the electron, so its velocity will be less than the electron for the same energy. Think of the proton as a truck and the electron as a Volkswagen. In an accident, a slowly moving heavy truck does just as much damage as a rapidly moving Volkswagen.

Your television set produces about 20,000 volts inside. The electrons

Figure 2-5 Surrounding Edward Lofgren (center), head of the Bevatron, are the four men who discovered the antiproton (left to right): Emilio Segrè, Clyde Wiegard, Owen Chamberlain, and Thomas Ypsilantis. (*Courtesy Lawrence Berkeley Laboratory*)

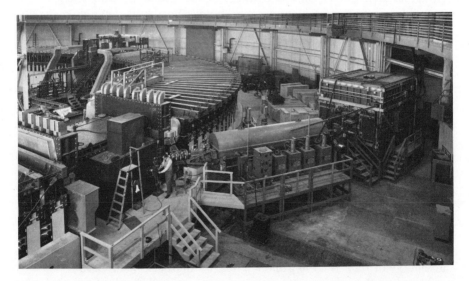

Figure 2-6 Part of the Bevatron, the proton accelerator at the University of California at Berkeley used to discover the antiproton. (*Courtesy Lawrence Berkeley Laboratory*)

in the TV tube therefore reach 20 keV (20 kilo-electron volts). They have enough energy when they strike the back of the screen to make the phosphor glow. A million-volt machine can accelerate electrons (or protons) up to energies of 1 MeV (1 million electron volts). At 1 MeV, an electron is moving at 94 percent the speed of light, while the heavier proton is only moving at one twentieth the speed of light. The proton, remember, is 1,836 times heavier than the electron. It doesn't have to move as fast to have the same energy as the electron.

To get energies greater than a few million electron volts with just an electric field is difficult, because high voltages have a tendency to leak off into the air. However, once an electron or proton beam has been set moving using electrical fields, it is possible to send radio waves traveling along in the same direction as the beam of particles. If the radio waves are properly tuned, the charged particles can gain energy from them just as a surfer gains energy from an ocean wave. Energies of thousands of millions of volts, or giga-electron volts (GeV), have been reached using this radio-surfing method. (This raises a side issue, a small matter of words. In the early days of particle physics, the unit for thousands of millions of volts was called a billion electron volt or BeV. That's where the Berkeley Bevatron got its name. However, a "billion" means different things to different people. In the United States, for example, and in this book, a billion means a *thousand* million. In Great Britain and some other countries, a billion means a *million* million. Americans call a million million a *trillion*. Particle physics is truly international today, unlike in the 1940s and 1950s. Physicists around the world have set up international standards for names. By common agreement, billion is now *giga* and BeV has been replaced with GeV. Trillion is now called *tera*.)

Billions (or to be more precise, gigas) of electron volts are no longer the limit in particle accelerators. Fermilab in the United States is now operating a superconducting proton accelerator designed to produce protons with an energy of a million-million electron volts or a tera-electron-volt (TeV) machine. The Fermilab machine is aptly called the Tevatron.

Physicists knew that the antiproton could only be created in a pair with a proton. Dirac's equations had shown that nature required a symmetry of charge when creating electrically charged matter from energy. This in turn told scientists how much energy would be needed to create antiprotons. A proton (or antiproton) has about 938 MeV of mass-energy. The pair would equal 1.8 GeV of energy. Even that would not be enough, though. The best way to turn energy into matter in a laboratory is to slam a high-energy particle into another particle—a proton, in this case, against another proton contained in a target. Now we're talking ideally about *four* particles following the collision: the two original particles (two protons) plus the proton and antiproton created by the energy of the collision. All four of these particles will come flying out of the collision with kinetic energy, or energy of motion. That kinetic energy will equal about one GeV for each particle. So—we have about two GeV of energy for the proton-antiproton creation, plus about four GeV of kinetic energy, which equals about six GeV of energy needed to create just one antiproton. (In real life, of course, many other subatomic particles come flying out of the collision, so it was important to keep a very close eye on the collision events in order to detect the antiproton.)

The Berkeley Bevatron, built in large part to find antiprotons, had enough energy to do the job. However, having a proton accelerator powerful enough was not sufficient. Segrè, Chamberlain, and the others would also need the right kinds of detecting instruments to see the antiproton when it was created. And they'd need those instruments arranged in such a way that they could distinguish the antiproton from other negatively charged massive particles that might also come spraying out of the collision.

Wiegand and Chamberlain had previously collaborated on other experiments, and they worked well together. They were already busy cooking up ideas for ways to find the antiproton. Segrè was interested, too, and felt that the best detector would be photographic emulsion. Physicists had been using photographic plates to detect and characterize subatomic particles for years. The method was simple. Take a stack of photographic plates and set them behind the target you were about to bombard with the electron or proton beam from your accelerator. The electrons or protons would smash into the target (usually a thin film of some metal like beryllium) and break apart some atoms in it. Those fragments would spray out of the target and into and through the stack of photographic plates, leaving tracks in the plates. In essence, the subatomic fragments took their own pictures. Positioning sets of magnets around the plate stack would bend the paths of electrically charged particles as they passed through the plates. Also, the more mass and charge the particle had, the thicker the path it would make. By measuring the thickness of the paths, the degree of curvature of the paths, and other properties, physicists could determine the nature of the particles that sprayed through the stack of photographic plates (Figure 2–7).

Segrè preferred photographic emulsions as a detector. He was the team leader and would have the final say on the makeup of the experiment. He was also an extremely cautious man; he had a formidable professional reputation that he wanted to protect. It would have been horrible to have claimed discovery of the antiproton, only to have the "discovery" shown to be fallacious because of faulty detectors, erroneous data, or some other glitch. Segrè perhaps wanted photographic plate detectors because he was familiar with them. And of course, their reliability was a known factor.

Owen Chamberlain, however, had come across another kind of detector that he felt would do the trick and still satisfy Segrè: a *velocity-selective Cerenkov counter*. The Cerenkov counter uses the Cerenkov effect, first predicted in 1934 by Russian physicist Pavel Cerenkov.

When light passes through water, glass, or other dense transparent substances, it is ordinarily slowed down in proportion to that substance's *index of refraction*. For example, the index of refraction of lead crystal is 2. That means light travels through it at half of its speed in a vacuum. The speed of

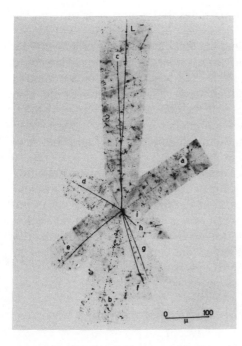

Figure 2-7 This photograph shows the existence of the antiproton. An antiproton (L) enters at the top of the picture and slows down in the photoemulsion. When it comes to a rest, it annihilates with a proton in either a bromine or a silver nucleus. Visible prongs of the annihilation "star" are mesons (a and b); a proton (c); and other protons and alpha particles (d to i). *(Courtesy Lawrence Berkeley Laboratory)*

light in a vacuum is 299,792 kilometers per second. In lead crystal, though, it is merely 149,896 kilometers per second. Nevertheless, a high-energy subatomic particle can pass through the lead crystal at a velocity faster than the official speed of light *for the crystal*. When this happens, the interaction of the particle's charge with the electrons in the crystal's atoms causes those atoms to emit light. The effect is similar to that of the shock waves a jet plane makes in the air when it exceeds the sound barrier, causing us to hear a sonic boom. In the case of Cerenkov radiation, we see radiation. The blue glow seen coming from used nuclear power plant fuel rods stored in cooling pools is Cerenkov radiation.

The *angle* of the Cerenkov radiation compared to the path of the particle depends on the particle's velocity. By measuring the angle at which the Cerenkov radiation is emitted, a scientist can determine the velocity of the particle causing the emission.

The detector Chamberlain had found out about used these effects. It detected Cerenkov radiation from particles traveling at certain selected velocities. Just as an electrical band-pass filter selects only certain specific radio signals, so this detector would allow only particles moving at certain velocities to pass through it and be detected. This kind of Cerenkov detector filter would work best if the incoming beam of particles were straight and tight. That meant having a way to focus the particle beam. With the help of physicist Orestes Piccione, the team devised a sophisticated magnetic lens that sufficiently focused the beam of particles that would pass through the Cerenkov detector.

Plate stacks and Cerenkov counters weren't the only detectors available to Segrè, Chamberlain, and the other two members of the team. They could also make use of detectors called *particle counters*, which literally counted the subatomic particles passing though them. The detectors would also trigger an electronic clock that would note the time when the particle passed through the detector. A pair of particle counters some distance apart could be used as a "stopwatch" to measure the particle's "time of flight" between the two counters.

The final experiment used a series of instruments including two bending magnets, two magnetic lenses, three scintillation time-of-flight counters, and two Cerenkov counters. The distance from the first bending magnet to the final scintillation counter was 80 feet. It was a special kind of maze that the antiproton mouse would have to traverse. The reward for successfully making the 80-foot trip would not go to the antiproton, though, but to Wiegand, Chamberlain, Segrè, and Ypsilantis.

Tracking Down the Wily Antiproton

The path of a particle in a magnetic field depends on the particle's *momentum*, which is equal to its mass times its velocity. Particles with the same momentum travel the same path in a magnetic field. The only particles that would emerge from the experiment's first bending magnet would be ones with identical momentum. However, they wouldn't necessarily be identical particles. Some could have a mass of, say, 2 and a velocity of 6, while others might have a mass of 4 and a velocity of 3. Both would have momentum of 12, since 2 times 6 is the same as 4 times 3. In the case of the antiproton trapping experiment, the spray of particles coming out of the target included both antiprotons and other kinds of subatomic particles called *mesons*. Mesons, like antiprotons, have a negative electrical charge, like the antiproton. They also emerged from the collision with the same momentum. It later turned out that there were 40,000 particles called *pi-mesons* for every antiproton. At the time, though, the researchers only knew that they needed a way to tell the mesons from the antiprotons.

One way to do that was by timing the velocity of the particles. Mesons are lighter than antiprotons, and so would be traveling faster. By timing the passage of the particles between two fixed points using the particle detectors, the experimenters could distinguish the slower antiprotons from the faster mesons. That would make it possible to compute the mass of the antiprotons (and the mesons, but their mass was already known), since the mass is equal to the momentum divided by the velocity.

Figure 2–8 is a drawing of the antiproton maze. First, the proton beam from the Bevatron hit a target (for their experiments the team used copper foil). Emerging from the collisions in the target was a spray of charged

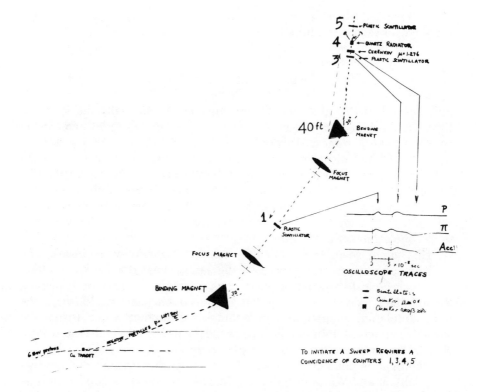

Figure 2-8 The maze of lenses and particle counters used to detect the antiproton at the Bevatron in 1955 was complex and ingenious. It eliminated any false readings, and assured the experimenters that the particles detected were indeed the sought-after antiproton. From beginning to end the antiproton-catching maze was 80 feet long. (*Courtesy Lawrence Berkeley Laboratory*)

particles. The first bending magnet extracted from the spray only those particles with a negative charge and the momentum predicted for the antiproton.

Then it was on to the next part of the maze. Here the particles passed through the first magnetic lens, which focused the beam of negatively charged particles into a tight stream. The focused beam then passed through the first detector. This was a *scintillation counter*, essentially a piece of plastic that would emit flashes of light when charged particles passed through it. It acted as the thumb on a stopwatch, starting the electronic clock as the particles passed through it. Then the particles passed through a second focusing lens and a second bending magnet. By forcing the particles to travel through two different bends the researchers made sure they eliminated all with the wrong momentum. Exactly 40 feet down the maze was a second scintillation counter—one that stopped the electronic stopwatch. In this way the researchers figured out the time it took for the particles to travel the 40 feet.

This "time-of-flight" technique allowed the scientists to determine the mass of the particles passing through the two scintillation counters. The mesons, with their lighter mass, were moving at practically the speed of light. The heavier antiprotons were traveling at only 75 to 78 percent the speed of light. The mesons would take about 40 billionths of a second to travel the 40-foot distance from the first to the second counter, while the antiprotons would take as much as 51 billionths of a second to make the trip. The difference was just 11 billionths of a second. However, the stopwatch pair of counters was so accurate that it could measure differences of one billionth of a second; discriminating between the two particles was well within the accuracy of the apparatus.

This might seem to be enough—but the team knew it wasn't. The collision of the proton beam with the target created a veritable flood of particles, most of which were mesons. So many mesons were passing through their special speed trap that at times one meson would trigger the first scintillation counter, and by coincidence another would trigger the second counter 51 billionths of a second later. To eliminate these false signals they put an "anticoincidence circuit" in the maze following the two scintillation counters. This was the fused-quartz Cerenkov counter that Chamberlain and Wiegand built. The Cerenkov counter responded only to particles traveling at somewhere between 75 to 78 percent of the speed of light. Those would only be antiprotons.

To be absolutely sure they would eliminate spurious signals, the team used two final "filters." The first was another Cerenkov counter in front of the first one. This counter would signal the passage of particles traveling *faster* than 78 percent of light speed. If its warning signal coincided with a signal from the quartz Cerenkov counter, or with signals from the stopwatches, that signal would be thrown out as false. The final filter was a third scintillation counter, set 80 feet from the beginning of the antiproton maze. This would record only those particles coming down the line from the beam's direction. Since the mesons had a short lifetime, most of them would have decayed before they had traveled 80 feet. The antiprotons had an infinite lifetime (provided they avoided contact with normal matter) and all of them would travel the 80 feet.

The entire collection of magnets and counters added up to a formidable maze, indeed. To be counted as a real antiproton, a particle would have to (1) get through the two bending magnets, and thus have the right momentum and charge; (2) travel the 40 feet between the two stopwatch scintillation counters in 51 billionths of a second, meaning it probably had the right mass; (3) register in the quartz Cerenkov counter with a speed of 75 to 78 percent light speed, meaning it definitely had the right mass, but (4) *not* trigger the warning Cerenkov counter; and (5) pass through the third scintillation counter, 80 feet from the beginning of the maze.

The Antiproton Game

The search for antiprotons began at UC Berkeley in early 1954. It took nearly 18 months to get everything set up and ready to roll. And when they had finished their preparations, the team almost ended up botching the job. Chamberlain, Segrè, and the others thought they had set their detectors to register the presence of particles that had the mass of protons. It turned out they were about five percent off. If they had started up the Bevatron's proton beam and turned on their detectors, they would have found nothing—even if floods of antiprotons had been passing through the detectors. Almost by chance, another scientist at the lab casually asked them, had they lined up their apparatus on protons? Well, sure, the team replied, but we'll check to be certain. They took a few hours to reverse the magnets in their apparatus, and ran a Bevatron proton beam through it—and found out the apparatus was not lined up to pass particles with the mass of the proton! Wiegand and Ypsilantis retuned the machines correctly. Then the Bevatron was started.

It was Wednesday, October 4, 1955. The front page of the *New York Times* that morning carried a story headlined "DRIVE TO HARNESS H-BOMB FOR PEACE IS ON IN 5 CENTERS; A.E.C. Head Says 20 Years Is 'Fair Guess' of Time Required for Success." According to the article, AEC Chairman Lewis Strauss was predicting that controlled nuclear fusion would be a reality by 1975.

The big news, though, was not about science. The Brooklyn Dodgers were locked in mighty struggle with the New York Yankees for the world championship of professional baseball. Leading the series three games to two, the Dodgers had gone into Yankee Stadium the day before fully expecting to win their first Series. It was not to be. The Yankees had shelled the Bums with five runs in the first inning and gone on to win the game 5 to 1. The rout was splashed all over page one, cheek by jowl with the fusion prediction. The Series was even now, three games to three. The two great rivals of baseball were about to meet again for the deciding game. In four previous encounters, the Dodgers had come away losers.

That morning, on the other side of the country, Clyde Wiegand set up a blackboard in the corner of the Bevatron building. Down at the bottom he was keeping track of the Series game score. On the rest of the blackboard he kept track of another score. The Bevatron started up. Protons began flying. In the first fifteen minutes of the experiment the detectors registered a signal. One signal. One antiproton. But it was a signal. It was real. Segrè, Chamberlain, Wiegand, and Ypsilantis had detected the first new particle of mirror matter in twenty-two years (Table 2–1).

That afternoon the brilliant shutout pitching of a youngster named Johnny Podres led the Brooklyn Dodgers to a 2–0 victory over the Yankees—their first World Series victory. After seven losing trips to the October classic—four of them with their crosstown American League rivals—the Bums from Brooklyn had finally won. On the next day's front page of the *New York Times* was the now-classic picture of catcher Roy Campanella hoisting a joyous Podres to the sky.

Table 2-1

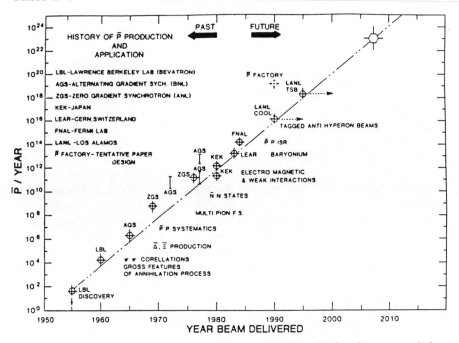

The history of antiproton production began in 1955, with the discovery of the antiproton at the Lawrence Berkeley Laboratory in California. More than 30 years later, copious amounts of antiprotons are created at laboratories around the world, and used as engineering tools to create new breeds of subatomic particles and to probe the microworld.

There was nothing about antiprotons, even though the most powerful atom smasher in the world had just hit the physics version of a home run. The news was known to the team at the lab, to Ernest Lawrence, and to a few others, but this was long before the days of instant science journalism. Segrè, Chamberlain, Wiegand, and Ypsilantis first would write up the results of their successful search in a paper to be submitted to *Physical Review Letters*, the premier journal of physics. They deliberately put pressure on the journal's editors to publish the report as soon as possible. They knew others were in the hunt for the antiproton, and they wanted to make sure their report was published first. It may have been their lobbying, or it may simply have been the momentous nature of their results: *Physical Review Letters* published their report of the discovery of the antiproton within two weeks of its submission.

The world finally heard of it on October 19, 1955. There it was, on the front page of the *New York Times*: "New Atom Particle Found; Termed a Negative Proton. 'Nuclear Ghost' Is Created From Energy in Coast Atom-Machine." People who only a few days earlier had cheered the Dodgers now puzzled over the meaning of "nuclear ghost" and "antiproton."

The scientific community was not puzzled. They recognized the particle's discovery for what it was, a confirmation of Dirac's work of nearly 30 years earlier, and further proof of the essential symmetry of nature. Science fiction writers were not too puzzled, either. The discovery of the antiproton simply confirmed what SF writers and readers had known all along—that the universe was a wonderfully strange place to live, and that science would continue to supply them with plenty of material to work with. Four years later, Chamberlain and Segrè would receive the Nobel Prize for the discovery. That same year the Dodgers, now in Los Angeles, would win their second World Series.

The discovery of the antiproton meant that two of the three essential constituents of the atom had known antiparticles. Still undetected was the antineutron. This would be a difficult beast to detect. The positron and antiproton were charged particles; scientists could detect them by searching for particles with the mass of an electron and proton but with opposite charges. The neutron, though, had no electric charge. Its mirror particle would not have any charge either. Instead, its magnetic moment would be the reverse of the neutron's.

However, these difficulties were quickly overcome. Eleven months after the discovery of the antiproton, a second team of Berkeley Radiation Laboratory researchers announced the discovery of the antineutron. Bruce Cork, Oreste Piccione, William Wenzel, and Glen Lambertson also used the Bevatron to ferret out this latest elusive "nuclear ghost" (Figure 2-9). This

Figure 2-9 The scientists who discovered the antineutron in 1956 are shown working on one of the Bevatron's giant magnets: (left to right) William Wenzel, Bruce Cork, Glen Lambertson, and Oreste Piccione. *(Courtesy Lawrence Berkeley Laboratory)*

discovery, too, made the *New York Times*, on Saturday, September 15, 1956. "Atomic Scientists Find 4th Particle. 'Anti-Neutron' Action Tops H-Bomb in Energy" read the tiny headline on the small article on page one. Most of the story was jumped to page 15. Four columns over from it was a tiny news item headed "Atom Program Pushed." The Atomic Energy Commission, the blurb noted, was "trying to encourage industrial companies to undertake 'chemical processing required for preparation of fuel materials for the peaceful atom program.'"

The Antiproton Tragedy

In 1959 Chamberlain and Segrè won the Nobel Prize in Physics for the discovery of the antiproton. They were surprised and delighted to receive it—surprised, because they never really thought their work had been of Nobel stature, since the existence of the antiproton was practically a foregone conclusion after the discovery of the antielectron; delighted, because they did get it. Personally and privately, they knew their work had been important, and the discovery of the antiproton was surely significant. However, they also knew it hadn't taken a surpassing intellectual effort to find it—just a powerful enough particle accelerator, and the right choice and arrangement of instruments to detect it. It was that last fact that was the cause of bitter feelings from one member of the discovery team, feelings that have never receded.

Of the four people on the team—Chamberlain, Segrè, Ypsilantis, and Wiegand—only Segrè and Chamberlain received the Nobel Prize. Tom Ypsilantis had worked hard on the project and was properly listed as an author on the discovery paper. But he also had less direct responsibility for the project than the other three. It was not too surprising he was not included on the Nobel list.

That Clyde Wiegand, on the other had, was left off the awards list remains a puzzle to this day. Some people feel that he should have shared the Nobel Prize with Chamberlain and Segrè. Segrè had led the team, had fought for and won the time needed on the Bevatron, and had brilliantly managed the entire effort. The basic idea of hunting down the antiproton, and the specific way to do it, were Chamberlain's. He also did many of the mathematical calculations that led to the experiment's success.

Wiegand, though, actually built the experimental apparatus. He was the one who had put the various detectors together, who had adjusted and readjusted the Bevatron's energy output. Wiegand had been involved in the effort from the beginning, and had played a central role in its success. Yet he was not given a share of the Nobel Prize. It is possible he was left out because he was not a faculty member at the University of California, but "merely" a member of the laboratory's staff. It could have been a case of someone being ignored by the "old boy network."

This kind of oversight is not unique, of course. Many notable researchers have never been recognized by the Nobel Committee for the significant work they have done. For example, John Wheeler, the great pioneer of black hole astrophysics (and the man who coined the name "black hole"), has never received a Nobel Prize. Neither has Wolfgang Panofsky, who discovered the neutral pi-meson particle, although many physicists consider that discovery to be at least as important as that of the antiproton.

Clyde Wiegand has never made a secret of his feelings about being left out. He still maintains that the detection of the antiproton would not have been made by the team were it not for his efforts—a claim almost certainly justified—and that he should have gotten a share of the Nobel Prize along with Segre and Chamberlain. Sadly, Wiegand's bitterness has made it difficult over the years for his colleagues to work with him. That is perhaps the one great tragedy of the discovery of the antiproton.

More Mirror Matter

The antineutron was not the last piece of mirror matter to be uncovered. In 1965 a team of researchers led by Leon Lederman of Columbia University created antideuterons. A deuteron, which is the nucleus of a variety or isotope of hydrogen called deuterium or heavy hydrogen, is a proton plus a neutron. (Normal hydrogen has only a proton nucleus.) An antideuteron, therefore, is one antiproton connected to a single antineutron. A third isotope of hydrogen, tritium, has a nucleus of one proton and two neutrons.

Lederman and his colleagues had built a very sensitive mass analyzer in order to detect the antideuteron. They installed it at the end of the Alternating Gradient Synchrotron (AGS), a proton accelerator at the Brookhaven National Laboratory on Long Island, New York. Then Lederman and his team bombarded a beryllium target with a beam of 30 GeV (30 billion electron volt) protons. Spraying out of the target came subatomic and atomic debris, including numerous mirror matter particles. Their mass analyzer detected about 200 antideuterons. A few weeks later an Italian team of physicists observed antideuterons using a 19 GeV proton beam from the Proton Synchrotron at the Center for Nuclear Research (CERN) in Switzerland. Lederman went on to become the head of the Fermi National Accelerator Laboratory (Fermilab) near Chicago. Fermilab is currently America's foremost creator and user of mirror matter.

Antitritium and antihelium have also been detected. Led by Yuri Antipov, a Soviet IHEP team bombarded an aluminum target with 70 GeV protons. They clearly detected five antihelium-3 nuclei among the 240 billion other fragments spewing out of the collision—a remarkable feat of physics. Normal matter helium-3 is a rare isotope of helium with two protons and one neutron. Antihelium-3, of course, has a nucleus of two antiprotons and one antineutron. (Standard normal matter helium, helium-4, has two protons and two neutrons.) Three years later, the same team produced four antitritium nuclei—antinuclei that have one antiproton and two antineutrons.

In 1978 physicists began using CERN's then-new SPS (Super Proton Synchrotron) accelerator to produce mirror matter. The SPS can slam protons together at 200 GeV of energy. One such experiment produced copious amounts of antideuterons, 94 antihelium-3 nuclei, and 99 antitritium nuclei.

This leads to an interesting extrapolation. At energies above 200 GeV, for every trillion antiprotons created and captured (which is about how many the CERN or Fermilab accelerators make per day) one could also capture about a hundred million antideuterons, 40,000 *antitritons* (the nuclei of antitritium), 10,000 antihelium-3 nuclei, and one antihelium-4 ("normal" antihelium) nucleus. However, even at 200 gigavolts of energy, it is not possible to reliably create more massive forms of mirror matter, such as antilithium (atomic number 3), antiberyllium (atomic number 4) or antiboron (atomic number 5). Those will have to wait for higher machine energies, greater beam currents and—especially—higher efficiencies for actually collecting and storing mirror matter.

Turn Left at the Moon: II

After dinner, Eric straightened up the living room while his father cleaned up the kitchen. His mother was tinkering with the Nissan. The uncertainty about his grandmother's health seriously disturbed him. He liked his "Granma Ginny" a lot—more, he was beginning to discover, than he had realized. He remembered visiting her and Grandpa John at their home in Atlantis Too on the ocean floor off California. They had flown out in a chopper to the small landing pad above the underwater community. The pad was just a speck on the otherwise clear ocean horizon visible from Santa Barbara. The oil well rigs from the last century were long gone, all underwater now. They had gone down to Atlantis Too in a slow elevator that changed the pressure as they descended to the large underwater village below. John had taken Eric out to see the underwater drilling rigs, and the new drill bit he was testing.

"It holds a little container of frozen antimatter that has enough energy to drill through a kilometer of hard rock," his grandfather had said. "It doesn't vaporize a large hole like a 'Trek Five' disintegrator ray, though. There are six antimatter beams that punch six tiny holes in the rock. That fractures the rock so the drill has no trouble grinding it up. The combination goes through hard rock like a hot knife through butter."

John had also shown him the compact power plant that generated the electricity for the oil rigs and the underwater homes of the miners. An antimatter-powered minisub had been hovering over the plant when they had driven by in Grandpa John's submersible. "They're bringing the monthly supply of antimatter out to the plant," he had commented. "It's expensive fuel—but at least there are no ashes to haul away. Your grandmother, of course, approves of that."

45

Grandma Ginny had later taken him on trips through the kelp beds and down into some deep canyons. She'd used the submersible's manips to snip off pieces of kelp, or to pick up oysters and rocks and then dump them into a basket. "I'll check these for pollution," Ginny had said at the time. "My last batch showed some oil pollution signs, which is strange. There are no oil rigs near this area."

They had then swooped down into a deep canyon. Eric's stomach felt funny as the submersible had flown over the edge of an underwater cliff and then zoomed down into the blackness below. Ginny turned on the floods.

"Aha!" she had yelled. They had stopped their drop and had moved back up the cliff a bit. She had spotted something. There, in the light, were black blobs of tar oozing out of a crack in the side of the cliff. "An oil seep," his grandmother had said with satisfaction. "Mystery solved. Well, I'll send a crew over here later and seal it up."

The phone beeped in the kitchen, startling Eric out of his reverie. His father picked it up. "Hello?" he said. "Ah! Hello, Mother Ginger! Sure. Hold on. KIRSTEN! Your mother!"

She came in from the garage, wiping her hands on her jeans. From Eric's end, the conversation was bewildering.

"Mother! How are you?

". . . Oh, dear! That's terrible news . . . .

". . . Oh! That's good news . . . !

". . . Oh! That's terrific! Yeah. Oh, wonderful. Wheww!

". . . Well, of course. We'd love to!

OK, I'll tell Jim and Eric, and call you back in a minute. Bye!"

She turned from the phone and hugged her husband. "Jim!"

"Well," he said, his voice muffled. "She sounded OK to me when I answered. And she must be feeling better if she called herself. What's up?"

"She went in this morning for a p-bar catscan of her head. They found nothing wrong with her ear, but they did find a small brain tumor. It had been pressing on the inner ear and causing the vertigo."

"Oh my God," Eric gasped. "Oh no."

"No, Red, it's OK." She turned and gave him a hug. "The thing was very small. So small that the doctor just turned up the current in the antiproton beam and cauterized the little sucker right then and there. She walked out of the hospital by herself. The first thing Dad had her do when she got home was call us."

"Great," said Jim. "And what are we gonna love to do?"

Kirsten grinned. "You two are going to love this. Mother said she had a lot of time to think while she had been lying in bed all these past weeks. She made a decision that, once she felt better, she was going to take time to do some things she's been postponing for too long.

"She's going to Mars."

"What?" said Eric.

"A two-week vacation tour."

"Awright. Rocky, man. I've always wanted to go to Mars."

"Well, Red, you can. She's invited us along. Your grandfather has to stay and check out a new oil field, so she wants the three of us to come along with her."

Eric lifted his arms above his head. "Rocky, man, rocky! So—when do we go? Wait'll Michelle hears about this!"

Four days later they were on a hypersonic flying over the North Pole.

The Boeing/Mitsubishi HT-303 Hyperplane was an antimatter-augmented hypersonic that flew from Paris to California in three hours. As they came into the Palmdale Interplanetary Spaceport, they could see the large dry lake at Edwards Air Force Base to the north.

"That's where the first American Space Shuttle landed," said Eric, pointing it out to his parents. He had been reading the history files his tutor had supplied him in preparation for the trip.

"Yep," said Jim. "I remember my grandfather telling me about watching it land that first time. And there—do you see that long line in the desert to the left of the dry lake? That's what's left of the first antiproton factory. It used a long superconducting linear accelerator. Ran for 20 klicks across the desert. Thirty years ago they used that machine to make enough antimatter for the first antimatter-powered rocket. A whole microgram." Kirsten chuckled at that. "We've got much better machines now," Jim continued. "They're a lot smaller and more efficient."

Eric turned from the window, pulled his pocket computer from his shirt pocket, and read some background data on the Palmdale Spaceport. There was a rumbling noise as the landing gear dropped down from the sleek underside of the hypersonic.

Chapter Three

Mirror Matter and Science Fiction

The discovery of the positron in 1932 by Carl Anderson of Caltech in Pasadena, California caused great excitement in the scientific world. But the excitement over the positron was not confined to the halls of academe. Several practitioners of a budding new genre of literature called "scientifiction" were entranced by the idea of positrons and antimatter. Their creative imaginations quickly conjured up visions of antiworlds, antistars, antigalaxies, and antiuniverses. Other science fiction writers used the idea of matter-antimatter annihilation as the energy source for galaxy-spanning starships that traveled faster than light. Those visions are the two main categories of mirror matter science fiction. However, a few science fiction writers have come up with other plot devices using antimatter.

In his famous "robot series" of stories, Isaac Asimov postulates the creation and use of robots with artificial brains as complex as those of humans. These artificial brains are laced with miniscule pathways that carry information in somewhat the same fashion as do the neurons of a human brain. The pathways are carved out by positrons. Asimov called them "positronic brains." In doing so he actually invented a new word: *positronic.* If you look up "positronic" in the authoritative *Oxford English Dictionary* you will find Asimov so credited. (He also invented the words "robotic" and "psychohistory.")

Another novel use of mirror matter is the central device of James Blish's 1954 story, "Beep." Blish cannibalizes Paul Dirac's original mathematical calculations to create an instantaneous signaling device using electrons and positrons. Blish's "dirac" device would never work in real life. Still, it makes a plausible-sounding explanation for the hoary science fiction cliché

Figure 3-1 Using mirror matter space tugs that look like long poles, astronauts have towed an asteroid into orbit around the Earth. A solar power satellite beams microwave power to Earth. Factories at its center melt and refine rock from the asteroid, and also power antimatter factories. *(NASA illustration)*

of talking on the phone to people on the other end of the galaxy like they were on the other end of town.

Seetee and Cannonball Express

Most mirror matter science fiction stories have been easily forgettable, proving the truth of Sturgeon's Law (named for the man who invented it, science fiction writer Theodore Sturgeon): "Ninety percent of everything is crap." A few mirror matter stories do stand out, however, and in the category of mirror matter planets, probably the best known are Jack Williamson's "seetee" stories.

Williamson's first "seetee" or CT (for *contra-terrene* or "anti-Earthlike") story appeared in the July 1942 issue of *Astounding Science Fiction* magazine, now known as *Analog Science Fiction/Science Fact*. The story, entitled "Collision Orbit," was published under the pen name "Will Stewart." One of the true grand masters of science fiction, Williamson has been writing since the mid-1930s. Williamson is still alive and well, living in New Mexico and still writing science fiction. His stories about planets made of mirror matter—"contra-terrene matter was what we called it then, with CT or 'seetee' being the abbreviation," says Williamson—were the first of their kind to appear in the field of science fiction.

Williamson first heard of mirror matter from *Astounding Science Fiction* editor John Campbell, the man who almost single-handedly reshaped the face of science fiction in the 1940s. "I went to Campbell with the notion of writing some stories about planetary engineers who would terraform new worlds," Williamson remembers. "He suggested that some of the worlds be contra-terrene. So these stories ended up being about the problems of dealing with worlds made of antimatter. I was working on the CT stories just before I went into the Army in 1942," adds Williamson. "I wrote the first one in 1941." The seetee stories turned out to have enduring popularity. "They've been reprinted several times since then. After the war I rewrote them into a novel entitled *Seetee Shock*. Then another serial of stories was published as *Seetee Ship*. They've been reprinted several times, by different publishers."

The seetee stories are not in the class of Hemingway or Bradbury. However, Williamson has never pretended they were. The stories are exactly what he intended them to be: unabashed adventures, featuring a main character named Rick Drake, a struggling asterite engineer who falls for the High Commissioner's daughter, Karen Hood. There are space-hardened men with names like Rob McGee and Paul Anders, and beautiful women such as Ann O'Banion and Jane Hardin. The stories abound with political maneuvering, space battles, and hopeless love crossing class barriers. There is stereotyped dialogue, with lines like "The rift was still untouchable as Karen Hood's bright hair," and "'Well, Captain. Shall we blow them out of space?'" and "'We aren't villains, Nick—certainly not all of us. That's what I've got to make you see!'" Still, the stories are fun to read, entertaining, and one of the first serious attempts at extrapolation into the future of a recent scientific discovery.

One fascinating sidelight to Williamson's seetee stories was their connection to the comics. Williamson used one of the stories, "Seetee Ship," as the basis for a comic strip called "Beyond Mars," which he scripted for the New York *Sunday News* in the 1940s. "Overall it didn't deal very extensively with antimatter," he admits, "but the first episode did have an antimatter theme to it."

Williamson wrote over a dozen seetee stories. Other SF writers have also played with the idea of mirror matter planets and galaxies. *A Wreath of Stars* by Bob Shaw is about an antineutrino world (not very likely in real life, but by definition an antimatter planet) that passes near Earth. In Frederic Brown's short story "Placet is a Crazy Place," the planet Placet orbits two stars. One of them is made of mirror matter.

A particularly well-worked-out mirror matter short story is "Flatlander," by Larry Niven. The story is part of Niven's Known Space series, and appears in his short story collection *Neutron Star*. "Flatlander" features his frequent main character Beowulf Shaeffer and an Earthman named Gregory Pelton, alias "Elephant," the "flatlander" of the title. Flatlanders are people born and raised on Earth, with physical reactions and mental attitudes conditioned by this upbringing. These reactions and attitudes are not always conducive to survival in space.

Elephant is rich, bored, and looking for adventure. He and Shaeffer end up journeying to a mysterious object barely within the borders of Known Space. All they know is that it is "unusual." Elephant (a miser) has refused to pay for the information about *why* it is unusual.

They almost find out the hard way. They fly a puppeteer-built spaceship (with its guaranteed indestructible hull) to their destination. Most of the rest of the "Flatlander" story details their increasingly strange experiences as they near the planet, which they dub Cannonball Express. The strange phenomena are clues to the planet's true nature, clues which Elephant and Shaeffer never quite understand. They almost get around to landing on it when their supposedly indestructible spaceship hull starts to dissolve around them. The two jump the remains of the ship back through hyperspace to a human-inhabited planet, where Elephant tells a puppeteer representative that he wants a refund for his ship, since the hull was supposed to be indestructible. The puppeteer asks for the circumstances of the hull's dissolution and Elephant tells their story.

In response, the puppeteer apologizes, but says it made "a natural mistake." It didn't think that antimatter was available anywhere in the galaxy, especially in that quantity.

"Antimatter?" whispers Elephant.

Of course, the puppeteer replies. Interstellar gas of normal matter had polished the planet's surface with miniscule explosions, had raised the temperature of the protosun beyond any rational estimate, and was causing a truly incredible radiation hazard. Humans are supposed to be highly curious—why didn't the two adventurers figure it out?

As for the ship's "indestructible hull," the puppeteer continues, though it is made of artificially strengthened molecules, it is nevertheless composed of normal matter. When enough of its atoms had been destroyed by annihilation from the impact of the mirror matter dust, the hull's molecules simply fell apart.

Elephant is stunned. He was about to try and land on a planet made of mirror matter. Beowulf Shaeffer gloats, and reminds Elephant of the one rule that almost all flatlanders forget in space: "If You Don't Understand It, It's Dangerous."

"Flatlander" is a fine example of so-called "hard" science fiction. Niven's story is grounded in real science. The strange phenomena Elephant and Shaeffer encounter as they approach Cannonball Express—the rising radiation count, the curiously polished appearance of the planet, the unusually hot protosun, and finally the destruction of their ship's hull—are what might happen if a mirror matter planet were plowing through a portion of our normal matter galaxy. And as Shaeffer says at the end of the story, "If we'd landed on your forsaken planet, we'd have gone up in pure light."

These stories by Williamson, Shaw, Brown, and Niven dealt with worlds made of contra-terrene matter—antimatter, or mirror matter. A 1943 short story by A.E. Van Vogt, "Storm," depicts a storm in space caused by a gas cloud made of ordinary matter coming into contact with a mirror matter gas cloud. Unlike the antimatter planet and star stories, this premise is closer to reality. But very few SF stories deal with mirror matter as it is actually used today—by physicists, in particle accelerators. One excellent example of such a book is *Broken Symmetries*, written by Paul Preuss in 1983. The story mostly takes place at TERAC, the particle beam accelerator built by the Japanese and Americans on the Hawaiian island of Oahu. TERAC—the Teravolt Accelerator Center—is the fictional version of the Superconducting Supercollider now being planned by the United States. It slams protons and antiprotons together at extreme energies. In the novel, some physicists have used the TERAC's colliding protons and antiprotons to discover a new kind of quark. There's some very serious political struggling going on between the Japanese and the Americans over TERAC. The Japanese, among other things, want to use the accelerator to produce large quantities of mirror matter (antiprotons, to be exact) for their planned experiments in mirror matter space propulsion. . . .

Obeying the Speed Limit

Of course, some science fiction writers do use the concept of mirror matter as an energy source for rocket propulsion. (Figure 3–2). The reason? Matter/mirror matter annihilation is the most powerful source of energy in the cosmos, and science fiction writers know they needed a lot of propulsive energy to get their characters from Earth to somewhere else in the universe. Science fiction stories about space travel and galactic empires often sprawl across countless millions of light-years of distance. However, the problem of getting from Earth to, say, the center of the galaxy in a few hours or days is a significant one. In fact, it cannot be done in the real world as we know it. The speed of light, about 300,000 kilometers (186,000 miles) per second, is the upper limiting velocity for the transmission of any kind of matter in the universe. Nothing can travel faster than light. Like the bumper sticker says,

Figure 3-2   Antimatter-powered luxury liners of the future could look like
this illustration of a nuclear rocket. At the left is the exhaust nozzle and mirror
matter combustion chamber for the propulsion system. Then comes a radiation
shield and the tank containing the mirror matter in electromagnetic suspension.
The rest of the ship includes tanks for liquid methane (the working fluid), and
near the front the crew and passenger modules. *(NASA illustration)*

"It's Not Just A Good Idea, It's THE LAW!" So nothing can get from Earth to
the galactic center in anything less than about 32,000 years. Two-week va-
cations to the fictional planet Trantor and back are simply not possible in
the real world.

Some science fiction writers have accepted this ultimate speed limit.
They still have their characters zip around the solar system, though, or even
head for the stars. Instead of mirror matter, they usually resort to propulsion
systems that range from chemical rockets to laser-pushed light sails.

A spaceship using chemical fuel and traveling at eleven kilometers per
second would take over 100,000 years to reach Proxima Centauri, the star
nearest to the Sun, 4.3 light-years distant. This "slow boat to China" theme of
space propulsion is best seen in stories featuring "space arks" or "generation
ships." Lawrence Manning's story "The Living Galaxy" (1934) and Don
Wilcox's "The Voyage That Lasted 600 Years" (1940) are probably the first
such science fiction stories. The two most influential ones, though, were
written by Robert Heinlein. His 1941 novellas "Universe" and "Common
Sense" were later combined into the 1963 novel *Orphans of the Sky*. They
are the archetypal "generation ship" stories. In a generation ship the original
crew of explorers will die, but their children will carry on. Since only the
first generation are true volunteers, we would want to make sure the slow

moving starship is a truly acceptable "worldship," with all the amenities and few of the problems of living on Earth.

One way to avoid generations of crew living and dying aboard a space ark is to put the crew into a state of suspended animation—a sort of human hibernation that would last tens or even thousands of years. Of course, no one knows how to do that for even a few weeks or months. Usually suspended animation is combined with a space propulsion system that reaches moderately fast velocities. A space ship traveling at about 3,000 kilometers per second, or one percent of the speed of light, would reach Proxima Centauri in about 430 years. Presumably it would be easier to use human hibernation for 430 years than for 100,000 years. One famous story that uses moderate interstellar velocity combined with hibernation is A.E. Van Vogt's "Far Centaurus."

Biologists are presently studying the aging process in cells and multicellular organisms. They are finding that our cells seem to be programmed to stop replicating after a given number of cycles. If they can find the right genetic switch, perhaps they can turn off the aging process and allow us to live the centuries that will be necessary to explore the stars using slowships powered by nuclear rockets. With death dead, our only enemies would be accidents and boredom.

One fascinating interstellar transport technique is the *Bussard interstellar ramjet*, invented back in 1960 by Robert W. Bussard, a research scientist in controlled fusion. The interstellar ramjet consists of a payload, a fusion reactor, and a large scoop to collect the hydrogen atoms in space. It carries no fuel because it uses the scoop to collect the hydrogen atoms that are known to exist in space. The hydrogen atoms are used as fuel in the fusion reactor, where the fusion energy is released and the energy fed back in some manner into the reaction products (usually helium atoms), providing the thrust for the vehicle. Many science fiction stories and novels have made use of the Bussard ramjet propulsion system. The most famous include the novel *Tau Zero*, by Poul Anderson and some of Larry Niven's "Known Space" stories.

Another "far-out" space propulsion system seen in science fiction is the *solar sail*. Light from the sun bounces off a large reflective sail surrounding the payload. The pressure from the light particles hitting the sail pushes it and the payload, a little like wind pushes the sails of an ocean-going sailing ship. Some of the better-known science fiction stories featuring solar sails are "The Lady Who Sailed the Soul," by Cordwainer Smith, "The Wind From The Sun," by Arthur C. Clarke, and Poul Anderson's novel *Orbit Unlimited*. Even more exotic is *laser-pushed lightsail propulsion*. Light from a powerful laser is substituted for sunlight, but the rest of the ship is like a solar sailship. The laser sailship, about as far from a rocket as possible, it consists of nothing but the payload and the lightweight sail structure. The rocket engine is the laser, energized by the sun. The propellant mass is the laser light itself. Laser-pushed lightsails have also found their way into science fiction. Their first use was in *The Mote in God's Eye*, a novel by Larry Niven and Jerry Pournelle. The "mote" in the title was a glaring green beam of light coming from the "eye" of a dark nebula formation that looked like the head of a god. The green beam of light was a green laser beam pushing a laser sail carrying an alien visitor. Niven and Pournelle got the idea for the laser-pushed sail from one of us (Robert L. Forward), who had already published a magazine article describing the technique. Forward hadn't thought much of the idea, because the laser only pushed outwards, so there seemed to be no way to stop. Forward later thought of a way and used it in his own novel, *The Flight of the Dragonfly*.

Breaking The Rules

Not all science fiction writers obey the real world's perpetually enforced upper cosmic speed limit. As often as not, they choose fiction over science. This is the reason for such common SF space opera conventions as "hyperdrive," "overdrive," "spacedrive," and "warp drive." All are simply fictional doubletalk methods of getting around the light-speed limit, so that galactic Empires and Federations become (at least fictionally) possible.

The explanations for them vary. In Piers Anthony's 1969 novel *Macroscope*, one of the characters explains hyperspace as being much like a wiggling line on a piece of paper. The one-dimensional line represents the four-dimensional space-time continuum. One can travel in a conventional manner, from one end of the line to the other. Or, says one character to another, one can jump from one wiggle to another one next to it, thus traversing a large distance in practically no time at all.

Another hyperdrive explanation was offered by science fiction writer Edmund Hamilton in his novel *Star Kings*. His interstellar drive avoided the speed of light limit by storing mass as energy. Hamilton was aware that as any material object approaches the speed of light its mass begins to increase. The greater the mass, the more energy is needed to move it closer to light speed. Hamilton thought the problem could be avoided by simply turning the increasing mass to energy, and using the energy for propulsion. Unfortunately, the universe doesn't work that way.

A third hyperdrive explanation is that of "the jump." Isaac Asimov uses this in his famous Foundation series of science fiction stories. The space traders and others simply "skip" across the galaxy, from one destination to another. They spend only a small amount of time in the real universe. The rest of the time the ship "jumps" through some other dimension or realm of space-time where the speed of light is not a limiting factor. Poul Anderson, in *Ensign Flandry*, has a more specific explanation for the "jump." It is, says one character, much like the instantaneous quantum jumps made by electrons as they shift from electron shell to shell around an atom. Somewhat the same idea is used by Clifford Simak in *Way Station*, and by John W. Campbell in his short stories "The Mightiest Machine" and "The Infinite Atom." The quantum jump idea is the best explanation for the transporter in *Star Trek*. In fact, SF writer James Blish uses just that explanation for the transporter in his *Star Trek* novel *Spock Must Die!*

The hyperdrive "jumps" of Asimov and Anderson often depend on the supposed existence of other dimensions of space-time, or on the existence of such realms as "hyperspace" or "subspace." One interesting version of this gambit is found in Murray Leinster's 1962 novel *Talents, Incorporated*. In this story, alien space ships get from one place to another by warping the space immediately around the ship into a different space-time continuum, one completely separate from our own. Different laws of nature prevail in this pocket continuum, and so the ships could happily travel at many times the speed of light.

In the end, such things as jumps, hyperdrives, and warp drives are likely to remain the stuff of science fiction. That includes the most famous warp drive of all—the one that powers Federation starship NCC–1701, the U.S.S. Enterprise.

Mirror Matter and Star Trek

All those warp factors are the product of the imagination of *Star Trek* creator Gene Roddenberry, a.k.a. "The Great Bird of the Galaxy." Roddenberry resorted to a based-on-fact fuel to provide the energy need to go where no man or woman had gone before. The Enterprise's faster-than-light engines are powered by mirror matter. That touch of fact is one of the reasons why *Star Trek* was—and is still—such a popular science fiction universe. Mirror matter as fuel for the Enterprise seems perfectly reasonable to us. Even Spock himself would find it so. It lends that sense of reality to the fictional universe of Klingons, Vulcans, and spaceships with all-officer crews. Of course, as immense as the energy released by matter/mirror matter conversion may be, it is still a finite amount of energy. That's not enough to propel anything faster than light.

Still and all, we can pretend; and the pretend universe of *Star Trek* is one of the most successful of science fiction. It has produced 77 live-action television episodes, 22 animated episodes, several full-length theatrical movies, at least two dozen original novels published by Pocket Books, and untold numbers of fan-generated short stories and novels. Mirror matter rarely figures prominently in the stories. It remains in the believable background most of the time, source of the energy that flings the Enterprise and its crew hither and yon across the galaxy. Occasionally, though, mirror matter takes prominence. Some examples include the television episodes "The Naked Time," "The Immunity Syndrome," and "That Which Survives;" the animated episode "One of Our Planets Is Missing;" and the movie *Star Trek III: The Search For Spock*.

"The Naked Time," a first-season *Star Trek* episode, has as its central plot device a molecule with rather strange properties: it causes a breakdown of psychological inhibitions. The crew of the Enterprise is accidentally infected and various people proceed to "act out" their basic natures. One crewman is a latent depressive and ends up killing himself. Sulu runs around the ship naked to the waist, pretending to be a swashbuckling swordsman. Nurse Chapel becomes very affectionate. Kirk gets excessively romantic about his ship. Spock breaks into tears. Even worse, engineer Kevin Riley declares himself captain, locks himself into the engineering bay, and turns off the matter/mirror matter engines.

The Enterprise begins spiraling down towards the planet below. Leonard "I'm A Doctor, Not A Crybaby" McCoy finally comes up with an antidote and everyone starts getting their emotions under control. The Enterprise, however, is still out of control. Matter and mirror matter cannot be mixed "cold," claims Scotty; the engines need 30 minutes of warm-up time before they can be turned on again. (Perhaps the Enterprise is a bit like a 1957 Chevy on a subzero day in Nome, Alaska?) In any case, the ship will hit the planet's atmosphere before the 30 minutes are up. Scotty finally saves the day by managing to "cold-start" the mirror matter engines.

Another mirror matter episode—one which stretches the bounds of credulity—is "The Immunity Syndrome." In this case the Vulcan-crewed starship Intrepid is eaten by a giant space-going amoeba made of negative matter. Negative matter is not the same as mirror matter. We know this because the Enterprise investigates and Kirk orders the ship to dive-bomb the thing. The crew is not destroyed in a flash of energy as they enter the amoeba. Instead, the engines begin working backwards (reverse thrust sends them deeper into the amoeba, for example) and the crew's life force is steadily drained from them. So it's negative matter. Logical, as Spock would say.

The crew nears death (and the episode nears the end of the hour) with Spock lost somewhere in the monster's protoplasm in a shuttlecraft. Kirk and Scotty and company come up with a plan to kill the amoeba with a mirror matter bomb. Scotty siphons some mirror matter out of the engine pods and into a magnetic bottle, and the Enterprise fires the bomb into the amoeba's nucleus. There it detonates and destroys the amoeba. Spock is rescued and quickly regains his usual rational health.

One of the closest looks we get at the Enterprise's mirror matter-powered engines comes in the episode "That Which Survives." A computer with a one-track mind is a central figure, guarding a planet whose inhabitants died long ago. A landing party (Kirk, McCoy, Sulu, and disposable ensign) beams down to the planet's surface. The computer projects a beautiful but deadly android woman onto the Enterprise, where it flings the Enterprise 990.7 light-years away, stranding the landing crew. Then the android returns to the planet and kills the disposable ensign. Then it returns to the Enterprise again (busy android, this), enters Engineering, and kills still another disposable crewman. Worse than that, it sabotages the matter/mirror matter integrator bypass control.

Scotty and Spock don't know about the sabotage. Spock calculates it will take the Enterprise 11 hours to return to the planet at Warp 8.4. Scotty has "a bad feelin'" about the engines. Thank heavens for his "feelin's." He discovers the sabotage to the matter/mirror matter integrator bypass control. In one of the more genuinely dramatic scenes of the TV series, Scotty wriggles up a tunnel with a special tool to try and undo the damage. Blue bolts of static electricity leap about him as he works; sweat drips from his brow; he has only seconds left before the mirror matter and matter intermingle and destroy the ship in a blast of annihilation energy. With Spock's help, Scotty repairs the bypass control. The ship gets back to the planet just in time for Spock and friends to save Kirk and friends from the beautiful but deadly android and its computer creator.

Besides the three seasons of live-action original *Star Trek*, one season of a *Star Trek* cartoon series was aired, using the voices of the original actors. Mirror matter figured in at least two cartoon episodes, including "One of Our Planets Is Missing." An extragalactic cloud creature made of energy and mirror matter enters our galaxy and begins eating planets. The Enterprise goes to investigate. The cloud captures the Enterprise and bombards the ship with mirror matter blobs, but the shields hold. The Enterprise then journeys through the cloud's "intestines" as antimatter villi suck energy from the ship. Spock uses a force field box to capture one of the villi, and Scotty uses it to jumpstart the Enterprise. Of course, all ends well. Spock mind-melds with the cloud (!!) and shows it that its planet-eating proclivities will kill millions of people. The cloud agrees to leave.

Finally, there are the *Star Trek* motion pictures. Mirror matter figures in all of them, of course, since the Enterprise is still being powered with matter/mirror matter engines. One deadly example is in *Star Trek II: The Wrath of Khan*. As all true *Star Trek* fans know, Mr. Spock is killed by the radiation from the Enterprise's leaking mirror matter converter. (Perhaps this scene should be mandatory viewing for all future mirror matter engine technicians.)

An even more dramatic example (in terms of special effects, anyway) comes near the end of *Star Trek III: The Search For Spock*. Kirk and his crew are trapped; the Klingons have them outgunned, outmaneuvered and outmanned (outpersoned? outklingoned? outentitied?). They want the Enterprise, and by God they're going to get it at last. But Kirk has one last card to play. He surrenders the ship and triggers a special computer command. Then he and his people quickly beam down to the surface of the Genesis Planet, leaving the Enterprise to the Klingon crew. They board the Enterprise. They hear the computer talking but they don't understand Federation-ese. The Klingon commander, aboard his ship, asks them to transmit the sounds. Apparently he did well in languages at Klingon U., because *he* understands Federation-ese, and what he hears is " . . . five, four, three, two, one. . . ." The ship's computer is counting down to self-destruction! The Klingon commander screams an order to his transporter room to beam the boarding crew off the Enterprise.

But it's too late. The Enterprise's computer orders the matter and mirror matter of the engines to intermix *in toto*. Down on the surface of the

Genesis Planet, Kirk and his crew watch the Enterprise explode and its fragments smash into the atmosphere. Shortly afterwards, Kirk kills the Klingon commander, rescues the reborn but mindless Spock, and tricks the remaining Klingon crew into beaming his people onto the Klingon ship. They successfully capture the Klingon vessel and Hikaru "I Can Fly A Goddamn Lunchbox" Sulu pilots it back to Vulcan. There Spock is reunited with his soul, which Leonard "I'm A Doctor, Not A Mystic" McCoy has been unwillingly sheltering for the entire movie. In this movie, as sometimes in real life, mysticism and cutting-edge science are mingled together.

Mirror Matter—No Longer Science Fiction

Since its discovery in the 1930s, mirror matter has mostly resided in the realm of science fiction. Positrons are a common byproduct of natural radioactive decay. And subatomic particles called mesons, 50 percent of which are mirror matter, occur in most energetic nuclear collisions. But it has become more and more apparent to astronomers and physicists that there are few naturally occurring *concentrations* of mirror matter in our solar system and even our galaxy. That has made mirror matter less attractive a plot device for science fiction writers. Most of the SF stories dealing with mirror matter were written in the 1930s, 1940s, and early 1950s. Larry Niven's story "Flatlander"and the *Star Trek* uses of mirror matter are the exceptions. Science fiction writers like to use plausible science for their stories. As Niven says: "Most of us science fiction writers like to write about black holes these days. Those look like they'll be easier to find than antimatter." The only recent example of a science fiction novel using mirror matter in any major way is *Fireball*, a first novel by British physicist Paul C.W. Davies published in 1987. In that story, Davies depicts worldwide disaster from an infall of meteorites made of mirror matter.

Most people today—including many scientists—still think of mirror matter as science fictional at best. But mirror matter is real. Physicists make and use mirror matter today, right now, in places as mundane as Switzerland, Long Island, and Illinois. There's nothing really mysterious about mirror matter itself, nor even about how it's made and used.

Mirror matter today is used at the frontier of particle physics. At several locations worldwide giant machines called *particle accelerators* create mirror matter year-round. The locations include Stanford University and the Lawrence Berkeley Laboratory in California; the Fermi National Accelerator Laboratory (Fermilab) in Illinois; the Brookhaven National Laboratory on Long Island, New York; the joint European multimachine facility (CERN) in Switzerland; and the Institute for High Energy Physics (IHEP) in the U.S.S.R. In some machines physicists make and collect beams of antiprotons. In others they make and use beams of positrons. These beams of mirror matter are then smashed either into beams of normal protons or electrons, or into foil targets of normal matter such as copper or beryllium. The collisions take place at enormous energies, and the results are sprays of exotic subatomic particles. Physicists read the electronic and photographic traces of the debris from these collisions the way fortune-tellers read palm lines. The difference is that the physicists are learning something real: They are learning about the essential nature and origin of the universe.

These particle accelerators are like gargantuan microscopes, allowing scientists to peer into the deepest recesses of atomic nuclei. What they see enables them to test new theories about the essential unity of all matter and energy. By using mirror matter, physicists have discovered the six different types of quarks and partially unified the four basic forces of nature (gravity, weak nuclear, strong nuclear, and electromagnetic).

Particle accelerators are also like time machines. The energies they use when smashing matter and mirror matter together have not existed in the universe since the first few moments of its creation some 18 billion years ago. What the particle physicists see emerging from those collisions are unified and semi-unified particles and forces that probably only existed during those few instants after the Big Bang. The greater the energies they use, and the more powerful the collisions of matter and mirror matter they can create, the more deeply into time and space they can look. So the particle physicists are on a continuing quest for ever-more powerful accelerators. The most powerful one these days is the Tevatron at Fermilab, near the town of Batavia, Illinois. The second most powerful accelerator is at CERN, outside of Geneva, Switzerland. The CERN machine will soon become the most powerful when a new addition is completed. Meanwhile, scientists at Stanford University are completing work on a machine called the LEP, or linear electron-positron collider machine. It will be the most powerful accelerator of its type in the world.

The prize for power will eventually go to the SSC, or Superconducting Super Collider, due to be completed in the later 1990s. One version of this accelerator will smash protons and antiprotons together at energies 20 times greater than the Tevatron. But even without the SSC, physicists are using mirror matter today, at this moment, as they probe the innermost secrets of the microworld. Mirror matter is science fact, and it is a vital tool in our continuing quest to understand the origin and nature of the cosmos.

Chapter Four

The Mirror Matter Mystery

The work of Dirac and the discoveries made by Anderson, Chamberlain, Segrè, Cork, Lederman, Antipov, and others, have shown that for every subatomic particle that exists there is (or can be) a corresponding antiparticle. The electron's mirror particle is the positron. The mirror particle to the proton is the antiproton, and the antineutron is the neutron's mirror particle. Neutrinos have their antineutrinos, mesons and lambdas their antimesons and antilambdas and so on. In fact, quarks, the most basic particles in the cosmos along with leptons (which include electrons and neutrinos) have their antiquarks.

All of this is true—in particle accelerators. When physicists smash atoms or subatomic particles together, they not only create new species of *matter* from the energies released, they also create the corresponding *mirror matter* particles. It was this clear result, repeated over and over again, that led science fiction writers to confidently postulate the existence of antimatter planets, stars, and galaxies. It only made sense. Various combinations of electrons, protons, and neutrons form atoms of matter, and configurations of atoms form molecules. In the same way, it is possible to envision various mirror particles forming mirror atoms, and mirror atoms forming mirror molecules. Such mirror atoms and mirror molecules would truly be mirror matter. In the more general sense, any antiparticle can be spoken of as antimatter or mirror matter. Indeed, there would seem to be no reason why mirror atoms, molecules, asteroids, comets, planets, stars, and galaxies should not exist. Mirror matter obeys all the laws of nature. As far as we know, gravity affects mirror matter to the same degree it affects matter. Chemical reactions involving mirror matter should be the same as those involving matter. There is nothing, it seems, to prevent mirror matter stars and solar systems from forming. Scientists as well as science fiction writers have thought this all makes sense, too.

Symmetry

One major reason they did is the concept of symmetry. The poet John Keats once wrote, "Beauty is truth, truth beauty." For many scientists this philosophical concept is more than just pretty words; it is a source of deep inspiration. It is a kind of article of faith that the universe prefers beauty and simplicity to ugliness and complexity. Paul Dirac once declared, for example, "It is more important to have beauty in one's equations than to have them fit experiment."

In fact, symmetry abounds in mathematics. The formula $E = mc^2$ is symmetrical. The terms on the left side really are the same as those on the right side. In physics, the so-called Conservation Laws are an example of symmetry. For example, one such law states that in an isolated system energy cannot be added or lost. Neither can the quantities of momentum and angular momentum. They are conserved. When James Clerk Maxwell mathematically united electricity and magnetism in the late eighteenth century, he added a term to make the equations symmetrical. That "arbitrary" mathematical term soon turned out to have cosmic reality. Maxwell's symmetrically beautiful equations led Henri Poincaré and Hendrik Lorentz to show there was a profound symmetry between space and time. Albert Einstein then showed that the mathematical space-time of Poincaré and Lorentz was real. He also showed that the speed of light, the velocity of Maxwell's electromagnetic waves, was the universe's ultimate speed limit.

Symmetry, then, is also a real property of the universe. Of course, that's easy to see. A ball of ice such as a hailstone has spherical symmetry: No matter how you rotate the hailstone, it looks the same. A snowflake has a hexagonal or six-sided symmetry: Rotate it through an angle of 60 degrees and it looks the same as it did before. A cubical crystal of salt has still another form of rotational symmetry. Then there's a form of symmetry we might call "reflection" symmetry; your left and right hands, for example, are symmetrical in this manner. In fact, we've already seen that antimatter is symmetrical with normal matter in a reflective sense.

We'll come back to the matter of symmetry a little later. Suffice to say at this point that the existence of symmetry, both mathematically and concretely, seemed a powerful argument for the existence of mirror matter in the universe. Dirac had first proposed the existence of mirror matter in order to satisfy the deep philosophical intuition that the universe *is* ultimately balanced, symmetrical. For every particle of matter there *should* exist a particle of mirror matter. Carl Anderson and Robert Millikan seemed to prove Dirac's intuition correct. If it is correct, then half the universe should be made of mirror matter.

Since then, physicists have observed this matter-mirror matter symmetry in the creation of mass from energy in particle accelerators. For every particle matter created a corresponding antiparticle is also created. Nevertheless, there is one serious problem with these arguments for the existence of mirror stars, mirror galaxies, and mirror planets: Observational facts do not seem to support their existence.

Where is the Mirror Matter?

Of course, mirror matter does exist outside of particle accelerators. The first particles of mirror matter detected were positrons created by cosmic rays. The mirror matter associated with cosmic rays is created by the energetic collisions of the cosmic rays with other normal matter. This mirror matter comes showering down from the upper atmosphere along with the other secondary particles created by cosmic ray collisions with atoms and molecules in the atmosphere. These cascades of particles and antiparticles hit molecules of air in the lower atmosphere and trigger still more showers of particles and antiparticles. The mirror matter produced in these cosmic ray cascades includes positrons. It was these positrons that Carl Anderson spotted in 1932.

In recent years, astronomers and physicists have attached special particle detectors to balloons that float into the upper reaches of the atmosphere. Earth-orbiting satellites such as HEAO–3 (High Energy Astronomical Observatory) also carry particle detectors, observing radiation and subatomic particles streaming past Earth from other places in the universe. These balloon- and satellite-borne detectors have spotted something unusual near the center of our own Milky Way galaxy. Inward near the galactic core is an object containing a lot of *positronium*, a pseudo-atom made of an electron circling a positron. Positronium soon decays and annihilates, of course, releasing two 511 keV energy gamma rays (this characteristic "signature" of the annihilation of positrons and electrons is how the positronium at the galactic center was spotted). What is unusual is that the positronium is still there, and so must be present in large amounts. Furthermore, the object's brightness varies over time, something like a variable star that swells and contracts. The reason is a mystery. The nature of the brightness variation has allowed astronomers to determine that the object involved is no bigger than a hundred trillion kilometers in diameter. Thus it is not a planet or a star; in any case, no planet or star could possibly be made of positronium. One possibility is that it is a cloud of gas some hundred trillion kilometers wide surrounding a black hole. A black hole could easily be a source of positrons and thus positronium. Also, the debris from a supernova explosion contains billions of tons of freshly minted radioactive isotopes. As they decay, they emit positrons. So it's conceivable the "positronium object" is really the positron-emitting debris of a supernova.

Nonetheless, we know there could not be large amounts of mirror matter in the universe. If there were, some of it would inevitably be coming into contact with normal matter. The result would be annihilation reactions. Electron and positron annihilation would have the radiation signature of 511 keV of gamma ray energy. The annihilation of nucleons (protons and neutrons) and antinucleons has a different signature. (Chamberlain, Segrè, Wiegand, and Ypsilantis knew this when they devised their experiment to detect antiprotons. As part of that experiment they had to discount the flood of mesons that were also created when protons slammed into the nuclei in the target.)

The annihilation of protons and antiprotons creates a form of meson called a pi-meson, or pion, made of a quark and antiquark. Pions almost immediately decay into gamma rays. The annihilation thus creates gamma rays as well as pions. This pattern gives proton-antiproton annihilation its own special "signature." Astronomers and physicists know these signatures well. But when they examine the universe, they find practically no signs of such signatures. The positronium object at the center of the galaxy is one of the very few exceptions. This apparent lack of mirror matter extends from our own solar system, throughout (most of) our galaxy and on out to the local group of galaxies and the super-group and super-galaxy cluster which contains us.

We can even say with confidence that at least some of the objects known as quasars are not made of mirror matter. Quasars lie billions of light-years from us. They are visible only because they shine with prodigious energies. Some astrophysicists once suggested that the energy of quasars might be caused by matter-mirror matter annihilation. However, we know now that there's nothing unusual about the quasars' radiation except the amount of it. The telltale signature of matter-mirror matter annihilation is missing. In fact, the best explanation these days for the energy production of quasars is the presence of black holes at their centers, sucking in matter and releasing enormous amounts of energy as they do so.

This apparent absence of the mirror matter radiation signature tells us two things. First, the universe seems to be missing mirror matter even in locations billions of light-years distant from us. Second, the universe was missing mirror matter billions of years in the past. The energy from quasars, traveling at the speed of light, left them billions of years ago. We see them not as they are now, but as they were only a few billion years after the creation of the cosmos, which in turn took place about 18 billion years ago.

Could there, however, be places in the universe that contain only objects made of mirror matter, places so separated from other parts of the universe that their mirror matter planets, stars, and galaxies never came in contact with normal matter planets and stars? A recent computer simulation seems to suggest there could not. I.W. Anaker and R. Kopelman from the University of Michigan used a Cray supercomputer to model a theoretical steady-state universe, one in which matter and mirror matter were continuously and spontaneously created from nothing. They wanted to see if such a universe would allow for the development of areas of matter and mirror matter separated from one another. In their simulation no such segregation developed. Matter and mirror matter tended to randomly diffuse throughout the cosmos and come into contact, causing annihilation. The real universe is not a steady-state cosmos. Still, the Cray steady-state simulation is an interesting one, producing results that seem to resemble the real universe's state of affairs.

Observations in astronomy and physics continue to show a remarkable absence of mirror matter from the universe. If the observations are indeed correct, what does it mean for the principle of symmetry? What it means is, things are not quite as simple as some scientists once believed. If the universe had been created with equal amounts of matter and mirror matter, then they would almost certainly have annihilated each other during the initial expansion of the cosmos. We should not even exist to be asking these questions. The universe, under such conditions, would contain nothing but energy. By the very fact that we are here, and that the universe manifestly contains matter as well as energy, we can theorize that the universe did not begin with equal amounts of matter and mirror matter.

Mirror Matter and Symmetry

Perhaps the reason that the universe did not begin thus may be that in a certain fundamental manner, the universe is not entirely symmetrical. When physicists talk about the symmetry of the universe they are usually referring to CPT, or charge, parity, and time, symmetry. Charge is electrical charge, positive versus negative. Parity has to do with handedness, left- versus right-handedness. Time has to do with temporal flow, past to future versus future to past. If the universe were indeed blessed with CPT symmetry then the following would be true: Consider an interaction between two positively charged particles, X and Y, which yields Z. Theoretically, if we reverse the charges (C), and run the interaction backwards (T) in a mirror-universe (P), everything would look exactly the same as it would in our universe. Of course, we can't do that, since we don't have access to another universe that is the mirror image of our own. However, it is possible to observe some reactions and see if they are the same under more limited conditions. And when that is done it turns out, much to the consternation of physicists, that the universe is not entirely symmetrical.

Certain subatomic reactions involving the interaction of particles are controlled by the so-called weak nuclear force, one of the four fundamental forces in the universe (the other three are the strong nuclear force, the electromagnetic force, and gravity). The weak force essentially governs the way some particles break apart and decay into others. The decay of a free neutron after about 15 minutes into a proton, an electron, and an electron antineutrino is governed by the weak force. So is the radioactive decay of some isotopes—for example, that of carbon-10 into boron-10, an antielectron (positron), and an electron neutrino. These are called beta decays, since the electron was once called the beta particle. They're also referred to as weak interactions.

In 1956 two physicists named Tsung-Dao Lee and Chen Ning Yang discovered that certain interactions involving the weak force are not completely symmetrical. It had long been assumed by physicists that whatever could happen in the real world would also happen if everything were reversed as if it were being observed in a mirror. This assumption of left-right symmetry, we've already learned, is the conservation of parity. Yang and

Figure 4-1   Professor John Cramer of the University of Washington in Seattle
has proposed a way to distinguish stars made of normal matter from those
made of mirror matter. *(Courtesy University of Washington)*

Lee found that this was not always true for weak interactions. Physicists
would say that parity invariance and charge invariance were violated.

Other scientists quickly confirmed Yang and Lee's work. In one experiment, beta radioactive atoms of cobalt-60 were cooled down to nearly absolute zero and subjected to a very strong magnetic field. The magnetic
field oriented the magnetic atoms of cobalt so their spin axes were all lined
up in the same direction. The cobalt-60 was emitting electrons (beta particles, remember?), since it was radioactive. The researchers discovered that
these electrons tended to be emitted in the same direction as the spin axis
of the cobalt-60 atoms. Suppose we looked at this process in a mirror that
reversed left and right—a "parity-reversing mirror." The spin of the atoms
would be reversed in the mirror, but the emission direction of the electrons
would not be reversed. The picture in the mirror *would not be the same as
the real world.* Parity is not conserved. So profound was this discovery by
Lee and Yang that they received the Nobel Prize in Physics just one year
later, in 1957.

The discovery of the violation of charge symmetry and parity symmetry
was most upsetting to many scientists. Some of them tried to regain their
own balance by claiming, essentially, "Well, maybe charge and parity can
be violated separately, even in the same interaction, but the combination
of C and P as CP is still symmetrical, still invariant." Unfortunately, this

also turned out to be untrue. In 1964 it was found that certain other weak interactions violated CP symmetry.

Ironically, this very lack of symmetry in the universe reveals a way to detect mirror matter stars and galaxies. It has to do with *helicity* and *polarization*. Helicity, as the word itself suggests, has to to with a helix shape—a spiral. Helicity in this case refers to the spin direction of positrons and photons. University of Washington physics professor John Cramer (Figure 4–1) first wrote about this method in *Physical Review Letters* back in 1977.

Mirror Matter, Polarization, and Helicity

The thermonuclear process that powers stars involves the fusion of light elements (like hydrogen nuclei, or protons) into heavier elements (like helium nuclei). In the process, protons get turned into neutrons, and some matter gets turned into energy ($E=mc^2$ again). This conversion of protons into neutrons takes place through the weak interaction process. It includes the release of anti-beta particles—positrons. When positrons are emitted in this fusion process, they tend to be in what Cramer has called a "right"-helicity state, their spins aligned along their lines of flight. This is all a result of that violation of parity first discovered by Yang and Lee. In a sense, positrons emitted in thermonuclear reactions tend to be "right-handed" when it comes to spin.

The positrons go flying out of the thermonuclear reactions, and zipping out through the star where the reactions take place. Charged particles like electrons and positrons produce a form of electromagnetic radiation when they travel along through matter. This radiation has the jaw-breaking name of *bremsstrahlung*, a German word that means "braking radiation"—radiation caused when the electrons and positrons are slowed down. This bremsstrahlung radiation will be polarized. The positrons in this case tend to have a right-helicity. That "right-handedness" gets transferred to the radiation, which ends up being right-circularly polarized.

Polarization has to do with the oscillation of light waves. Suppose you and a friend are holding the two ends of a jump rope. You give your wrist an up-and-down flick. That sends a wave traveling down the jump rope to the other end. The wave is oscillating in an up-and-down direction. You can speak of the jump rope wave as being *polarized* in that direction. If you'd flicked your wrist left to right, the wave in the jump rope would have a left-right polarization. As long as the wave in the jump rope is always in the same line, it is said to be polarized. The wave's direction of oscillation could also change randomly from one orientation to another. Suppose you and your friend could both flick your wrists at the same time, but in different directions. You wave the rope up and down, and your partner snaps it left and right. The 90 degree difference in the direction of the waves would result in a polarization of 45 degrees. Or, you and your friend might wave the rope in a totally random manner, causing many waves to travel up and down the jump rope in random directions. In that case you could say the jump rope is un*polarized.*

The same is true of light waves. Usually light is unpolarized; all the waves of light you see are vibrating in different orientations. Suppose, however, you put on that pair of polarizing sunglasses you bought at the drugstore. The plastic lenses of the sunglasses let through to your eyes only waves of light vibrating in one orientation. That means the light becomes polarized. This is a bit like what happens inside the star we've been talking about. The right-helicity of the positron's spin will cause the radiation it emits to have a right-handed polarization.

Of course, these positrons don't get very far before they hit a particle of normal matter. When that happens, the positrons and the normal matter particles annihilate one another, creating energy. Some, if not most, of that energy is in the form of *photons,* or light particles. Once again, the right-helicity of the positrons gets transferred to the photons. The photons become polarized in a preferentially right-spinning direction.

All of this happens in stars made of *normal* matter. For stars made of mirror matter, exactly the same processes will take place, but with one big difference: The thermonuclear process in antistars would turn *antiprotons* into *antineutrons.* That would produce both energy and *antipositrons.* Ah! But an antipositron is—an electron! Because of parity violation these electrons will not have their spins aligned parallel to their lines of flight. Rather, their spins will be *antiparallel.* The bremsstrahlung radiation they produce will be *left*-circularly polarized. So will the annihilation radiation they produce when they hit antiparticles in the antistar.

Thus the secondary energetic photons from matter stars and mirror matter stars will tend to be polarized in opposite directions. In principle, this means there is a way to distinguish between matter and mirror matter stars just by examining the light coming from them. No need for Larry Niven's fictional attempt to land on an antiplanet! The question now is: Can this radiation difference ever be observed in real life?

It's a good question. The thermonuclear processes that power stars (and presumably antistars) take place deep within their interiors. The radiation would have to be very "lucky" indeed to escape the star without first being absorbed within. So there is probably very little chance of simply examining the light from a normal antistar and deducing its (anti)nature.

However, not all stars act normally. When they are old and dying, in fact, some stars act very abnormally. They blow up, becoming supernovas. One such cataclysmic explosion took place in the Large Magellanic Cloud, 163,000 light-years away from us. The light from Supernova 1987a took 163,000 years to get here. It was first spotted on February 23, 1987, mere hours (relatively speaking) after the star actually exploded. Supernovas like 1987a offer the possibility of directly observing the matter-mirror matter helicity that John Cramer has written about. When a star blows itself up like this, we have a brief chance to look at the deep interior of the star. Also, a lot of the production of energy as the supernova explodes is taking place at the surface of the star and not just in its center. So it's possible to look at a supernova like 1987a and determine if it's a supernova or an antisupernova.

"Possible" is the operative word here. Cramer has estimated that the peak flux of polarized radiation from a supernova would take place about a month after the explosion. One way to detect it would be to use an instrument called a *polarimeter*. The polarimeter in this case would have to be about a meter (3.3 feet) in diameter, and situated in space. Scientists would need to start observing the supernova within two weeks of the explosion. They would have to watch it with the polarimeter for a full year in order to accumulate enough data to determine the radiation's helicity. That's a long time. More to the point, there is no one-meter polarimeter in space. In fact, no one has ever even *built* such a specially designed polarimeter. To design this instrument would be a challenging problem, but it is possible. A launch vehicle like the Space Shuttle could easily put it into orbit. So could the Hermes manned shuttle to be launched by the European Space Agency in the late 1990s.

It's all rather ironic. Supernova 1987a was the first nearby supernova observed since the invention of the telescope. The last supernova in or near our galaxy exploded in 1680 (Earth time). It is called Cassiopeia A, and it is about 9,700 light-years from us, in the direction of the constellation Cassiopeia. Galileo had invented the astronomical telescope more than 70 years earlier. For some reason, though, no one ever saw Cassiopeia A. Since then, astronomers had seen more than a hundred supernovae in other galaxies, making observations that have been very helpful. But even the nearest galaxies are nearly a million light-years distant. It is so much better to observe a supernova closer to us. So astronomers had been waiting for more than 200 years for this moment, and it has been a spectacular boon. They have been able to use all the high-tech instruments of late twentieth-century astronomy to examine the catastrophic event. They've even been able to use the new neutrino "telescopes" in Japan, the United States, and Switzerland to detect the flood of these strange "ghost particles" from the supernova.

But as far as *antimatter astronomy* is concerned, Supernova 1987a was *too close* to us! If only the star had been just ten or fifteen light-years further distant, its light would not have gotten here until the late 1990s. By then we would have been well into the creation of mirror matter technology. That in turn would have spurred the design and creation of polarimeters and other instruments that could detect Cramer's helicity difference. Now we have to wait some more, and hope the instruments we need will be ready when the next nearby supernova explodes.

Mirror Matter and Creation

So far no one has found an interaction in which the *total package* of charge/parity/time (CPT) is asymmetrical. However, Yang and Lee's discovery has shown that in some ways the universe is not entirely symmetrical.

This subtle distortion must have been woven into the fabric of the cosmos at its very beginning, during the first few instants of the Big Bang. At the level of weak interactions the universe is slightly asymmetrical. We only see this asymmetry at the microlevel of protons, neutrons, and electrons. On the larger macro-level of stars, galaxies, and superclusters, we see another lack of symmetry: the gross imbalance between matter and mirror matter. Certainly physicists and astronomers now suspect that the absence of mirror matter may also have its roots in the first instants of the universe.

One of the best explanations for the lack of mirror matter in the universe is also one of the simplest. At the very beginning of the universe more matter came into being than did antimatter. During the first billion-billion-billion-billionth of a second of the Big Bang, the temperature of the cosmos was a billion billion billion degrees. At that temperature matter and antimatter were created, but in a slight imbalance. For every billion particles of antimatter there were a billion and one particles of matter. In the tiny sub-proton-sized universe at the beginning of time, the matter and antimatter came into contact. Micro-moments later, all that was left was the tiny excess of normal matter. The universe kept expanding and cooling. Today, 18 billion years after those first fiery moments of the universe's birth, we see all around us the surviving remnants of that ultra-annihilation: the galaxies, the stars, the planets—and us.

At about the same time as the ultra-annihilation event, the One Great Force in the universe "crystallized" into the four forces we see today: gravity, electromagnetic, weak nuclear, and strong nuclear. The asymmetry in weak interactions became "frozen" into the pattern of the cosmos. Are the two asymmetries related? It is quite possible. But scientists still don't know for sure.

This very simple explanation for the missing mirror matter leaves only one question unanswered: *Why* was there more matter than mirror matter at the beginning of the universe? Astronomers, physicists, and philosophers have spent years searching for the answer to that question. The most recent answer comes from the blending of quantum physics with cosmology, the study of the origins of the cosmos. Quantum physics tells us that it is possible for matter (and mirror matter) to spontaneously form out of nothing—to appear suddenly in the vacuum of space. This is perfectly fine with the universe, as long as the matter and antimatter thus formed quickly *return* to nothing at all. The process is called *virtual pair formation*. "Virtual" in this case means temporary. "Pair formation" refers to the appearance of a mirror particle in the vacuum of space along with a normal matter particle. Extremely powerful gravitational fields can cause virtual pair formation. Black holes have incredibly powerful gravitational fields, and so virtual pair formation of matter and antimatter probably takes place near their edges. This may be the source of the positronium astronomers have found near the center of our galaxy.

Another cause of virtual pair formation is the so-called *vacuum fluctuation*. A vacuum, by definition, is the total absence of anything. A vacuum seems "smooth" and empty. However, that's because we "see" a vacuum with human eyes. Suppose we had an extremely powerful microscope, so powerful we could see things as small as a quark. Now let's use that microscope to look at the nothing of a perfect vacuum in deepest space.

What we would see at this level of magnification is astonishing: the actual quantum nature of space itself. At this level of reality, the vacuum seems to fluctuate back and forth, up and down, in and out. It looks "foamy," like the froth on beer or the sea-spume spun off the tops of waves in a storm. And in that foaminess, in that fluctuation, particles and antiparticles are born and annihilated in a flash of an instant. We never see them, because they come and go too quickly for that. But the mathematics of quantum physics says firmly that this bizarre state of affairs is real. The very fabric of the cosmos is like beer foam. Bubbles of quantum carbonation appear and pop from instant to instant, all beyond our ken.

What does all this have to do with mirror matter, and the beginning of the universe? Just this: Eighteen billion years ago, this huge universe we inhabit appeared. It appeared, apparently, out of literally nothing. The cosmos suddenly began expanding from a point of no-place and no-time. Not only did all matter and energy (which are the same thing in different garb) appear at the instant of the Big Bang—so did time and space themselves. How? Why? Some cosmologists today think the entire universe is itself the result of a sort of vacuum fluctuation. Only it wasn't a fluctuation in a vacuum, which really is something after all. Nor was it a fluctuation in space. Like time, space began with the Big Bang. No, the universe may have begun as a kind of vacuum fluctuation in the true nothingness beyond the universe.

Vacuum fluctuations are purely random in nature. You never know when or where they will happen. You never know what pair of particles and antiparticles will suddenly pop in and out of existence. The same was true, say some cosmologists, of the fluctuation which birthed the cosmos. It was random. The laws of nature in the new cosmos were random in their creation; they could have been different. And random also was the amount of matter and antimatter that condensed out of energy in the first few moments after the Big Bang. We and the universe, says this explanation, are merely a random space/time/matter/energy fluctuation in something that is beyond knowledge because it is nothing at all. The fluctuation has lasted 18 billion years. It may go on forever. It may eventually collapse back into the nothingness of its origin. No one is sure. The universal expansion may never stop. Or the cosmos may some day, hundreds of billions of years from now, begin collapsing back to an ultimate Big Crunch.

If it does, there will be plenty of mirror matter—at least for a few instants. The temperature of the now-proton-sized universe will again be a billion billion billion degrees. Once again, matter and mirror matter will form. But it will all be for naught. As the cosmos crushes down to the ultimate singularity, subatomic particles of all kinds and charges will simply dissolve into a quark soup. Then the quark soup will become energy.

And then?

Who knows? Perhaps the Big Crunch will "bounce" into a new Big Bang. Or perhaps the Big Crunch will just crunch back into nothingness. Then the words of Hotspur in *Henry IV, Part 1*, will come true, and "time will finally have a stop."

And so will mirror matter.

Turn Left at the Moon: III

E ric's grandparents met them in the terminal. They spent the rest of the day in their hotel room resting and packing for the long trip. They were also checked out by the Spaceport medical staff. Mars was still free of almost all communicable diseases (except the common cold), and the authorities wanted to keep it that way. There was some concern expressed about Ginny's tumor operation, and whether the problem might crop up again during the flight. The concern was not too serious, though. An antimatter catscan machine was standard equipment on the Mars Longliner—they had enough antimatter available from the fuel tanks to operate one. If Ginny's tumor made a sudden reappearance, they'd just zap it again.

Two hours before the flight they all took spacesickness pills. They said goodbye to Eric's grandfather (Ginny with a passionate kiss), boarded the transfer bus, and sped out across the field to the long runway that stretched across the California desert. Parked at

the end of the runway, ready for its flight into space, was a McDonnell-Douglas AS-3 Spaceclimber. The antimatter-powered plane would take them to the o'neill at L-4. There they would board the Mars Longliner for their vacation trip to the Red Planet.

Eric looked out the window of the bus as they approached the sleek spaceplane. It had the general shape of a pointed paper airplane, but with twin tails. Its surface was covered with gray-white insulating material that protected it on reentry. The spaceplane was actually smaller than the hypersonic they had taken from Orly to Palmdale. As they got closer, Eric noticed there were no wheels; he could see right under the spaceplane to the other side. He nudged his grandmother.

"Notice that?" he said. "It's floating in the air. The thing must use superconducting levitation landing gear. And there must be some superconducting magfield motors under the runway to hold it up. Pretty rocky, huh?"

"Very rocky," Ginny agreed. "Hey, did you cut your hair again?"

"Ahhh, c'mon."

"Very unrocky, dear."

The 60 passengers were carefully tucked into their seats according to size and weight. Eric and Ginny were put in a front row, and Eric got one of the windows. He slipped on the headset and tuned it to the spaceplane's cockpit channel.

The engines started at a low level, and the spaceplane began its long glide down the runway. The superconducting motors in the runway saved on fuel by catapulting them up to speed and into the air. The markings on the ground became a blur. The engines suddenly burst into a roar, and Eric was pushed back into his seat.

"One kilometer," said the soft voice of the cockpit computer in his ear. More faintly came the voices of the human crew, calling off green lights, confirming switches, talking with ground control. "Two kilometers . . . three . . . four . . . " Insulating shields slid over the windows as the acceleration increased. The AS-3 Spaceclimber shook slightly as it passed through the sound barrier. ". . . max Q . . . six kilometers . . . throttle up to 94 percent . . . eight kilometers . . . ten . . . rockets on." The sound of the engines changed as they switched from airbreathing to rocket mode.

"Fifteen kilometers . . . twenty . . . twenty-five . . . antimatter augment on." The engines' muffled roar increased as milligrams of antimatter were added to double and then triple the rockets' thrust.

The rockets stopped abruptly. They were in freefall. The window shields slid back. The captain turned the spaceplane upside down so everyone could look out the top ports at the curve of the Earth below. The sky had turned from blue to black.

"Welcome aboard," said the captain. "We'll be at L-4 in about 12 hours. One hour from now we will pass through geostationary orbit altitude. At that time we should pass pretty close to the big glasser that supplies energy to the west coast communities in North and South America. We should get a good view of it," she continued. "Its solar cell array is 30 kilometers wide and 300 long. It produces about ten million megawatts of power. That's ten thousand gigawatts, or 2,000 of the old-time Earth power plants. In case you're interested, there are four other glassers in orbit—one each for Europe, Russia, and the east coast of the Americas, and the one under construction over Asia."

An hour later the view from the top ports slid away to the left. The captain was rotating the spaceplane so the passengers could see the glasser from the top ports. "Hello again," came her cheery voice. "If you look through the top viewports you can see the SPS for the west coast. Now, look closely at the narrow gap in the middle of the glasser's solar cell array. You may be able to see a round disk a few kilometers in diameter. That sends multiple beams of microwaves to the various power receiving stations in the oceans or at isolated locations in the deserts. The new superconducting technology has made it possible to put the receiving sites far from the population centers.

"Now, if you have very good eyes, you may also see the antimatter factories in that gap. When the power companies don't need all the energy from a glasser, then what's left over is used to make antimatter. It's an economical way to store energy in a compact and lightweight form. Otherwise the solar energy would simply be lost, or wasted. Some of the antimatter gets shipped to Earth, but most of it stays out here where it gets used for space propulsion."

Eric watched closely as they glided past the gargantuan sunlit structure. His father had left his seat and floated over to them. "I can see the microwave dish," said Eric. "Where are the antimatter factories?"

"The two that look like doughnuts," his father replied. "The smaller spheres are for processing materials from the Moon, and from the Aten and Helin asteroids."

"Jim, you're blocking my view," said Ginny.

"Whoops. Sorry."

The Spaceclimber passed through GEO altitude and continued its climb into space. Lunch was served; then dinner. Eric played chess with his grandmother on her slate. Then they tried DeathRace 3000 on his. His grandmother slaughtered him. The captain had positioned the spaceplane so the Moon was visible by looking forward out of the upper viewing ports, while the Earth could be seen looking backward out of the ports. The o'neill at L-4 was getting closer. Eric explained to a young girl about his age sitting across the aisle that L-4 is a point along the orbit of the Moon that is the same distance from the Earth as the Moon, but 60 degrees ahead of the Moon in its orbit. Anything (said Eric in a knowledgeable tone that had the girl very impressed) that got placed at that point has a tendency to stay there. Like an o'neill or an asteroid like Helin.

The girl soon lost interest in lagrange points, and started asking more personal questions. Eric, suddenly embarrassed, backed off. Drowsiness from the effects of the spacesickness pill set in, and he napped.

He woke up as the craft approached Dyson City at L-4. The captain rotated the ship and fired the engines to bring them to a halt at the docking port.

Chapter Five

Antiproton Technology

In the late 1950s and early 1960s, machines like the Berkeley Bevatron and the Alternating Gradient Synchroton at the Brookhaven National Laboratory were the "creme de la creme" of high-energy particle physics. Today, the Bevatron and the AGS have been surpassed by particle accelerators of immense power and sophistication. For all their power, though, current accelerators still fall into one of two categories.

Today's tools of antiproton technology are either electron accelerators or proton accelerators, with an occasional ion accelerator. Electrons make excellent "bullets" to shoot at a target. They're easy to make, easy to accelerate, and are truly elementary particles with simple, well-known properties like mass, charge, and spin that are effectively point-like in size. Physicists know exactly the type of interactions the electron is capable of, making it much easier to unscramble the complicated information coming from the spray of particles that result when electrons strike a target. Anything unusual must be due to the target nuclei, since the electron is so simple.

Unfortunately, the electron is very light in mass. To get a lot of energy into the electron for bigger and better collisions, it's necessary to increase the velocity of the electron to near the speed of light. If you attempt to divert its trajectory when the electron is moving that fast, it radiates electromagnetic energy in the form of light and X rays and slows down. Therefore, while the older low-energy electron accelerators could be built to go around in a ring that would fit on a university campus, the higher energy electron accelerators either had to be built in a straight line that stretched across miles of land (the Stanford Linear Accelerator, for example) or had to be bent with such a low radius of curvature that the ring became so large it passed over international borders (as does the LEP in Switzerland).

Most of the laboratories interested in building large particle accelerators use protons instead of electrons. As "bullets" to probe target nuclei, protons have the disadvantage of being complex "bags" of quarks that make it more difficult to interpret the data. But a proton is also 1,836 times more massive than an electron, so it can carry the same energy as a relativistic electron, while moving more slowly and radiating much less energy. Proton accelerators can be built on tracts measured in acres rather than square miles. Even so, as the proton energies became higher and higher, the bending magnets needed to keep the protons moving in a circle had to become stronger and stronger. Fortunately, superconducting magnets became feasible and the machines could be boosted in energy while still using the same tunnels.

There is a difficulty with using the technique of shooting high-energy "bullets" at a fixed "target." Since the target is an atomic nucleus that contains protons and neutrons, we can think of the interaction as that of a proton fired at a proton (or a neutron). When the kinetic energy (that is, the energy of motion) of the bullet proton is less than the rest mass of the proton or neutron that it is hitting (less than 1 GeV), then the bullet proton weighs as much as the target proton. When the two collide, they divide the energy and momentum. They both move off together in the same direction that the bullet proton was traveling, so that half of the kinetic energy in the initial bullet is still in the form of kinetic energy in the resultant particles. The other half of the initial kinetic energy has been deposited in the two particles, hopefully doing things to them so they will emit some strange new Nobel-Prize winning particle.

A better result would be obtained if we built two proton accelerator machines, each with half the energy capability, and slammed the two protons head on. Since the two particles have the same mass, momentum, and energy, the head-on collision would leave them standing still. Thus, all of the energy would go into exciting the two particles and creating bizarre new quark combinations.

The situation gets worse as the energy of the bullet proton begins to exceed 1 GeV, which is roughly the energy equivalent of the mass of the target proton. At Fermilab, the 400 GeV Main Ring proton accelerator increases the energy and velocity of the bullet proton, until the Einstein equivalence of mass and energy makes the bullet proton weigh nearly 400 times as much as the target proton. The heavy proton smashes into the lighter proton like a runaway truck into a Volkswagen bug. There is a lot of energy pumped into the "bug" particle, but a great deal of energy is still left in the kinetic energy of motion of the bullet. The amount of useful energy transferred into Nobel Prize-winning particles under these conditions is no longer half of the initial energy, but only *the square root* of twice the energy. For the 400 GeV protons at Fermilab, only about 27 GeV of energy gets converted into new particles. Now it become more advantageous to build two machines and arrange to collide the protons head-on, so you get all the initial energy turned into new particles.

Head-on Collision

There are two ways to build a facility to collide two high-energy particles together. One is to build two separate particle accelerators, one accelerating particles clockwise and the other accelerating particles counterclockwise. Since the major cost of building a new particle accelerator is digging the tunnel to put it in, you can save some on costs by having them share the same tunnel. Then at one or more places you have the two beams in the two machines cross through each other. By focusing the beams down to a thin stream in the interaction region, and carefully aiming the beams, you can get enough head-on collisions to keep a small army of scientists busy for years with the resulting data.

Another way to build a collider facility is even cheaper. You only build one proton accelerator (or use the old one), and instead of building another ring, you build a factory to make antiprotons. You take the antiprotons (with their opposite charge) and put them into the accelerator ring along with the normal matter protons. Then you turn on the accelerator. The mirror matter protons have an opposite charge to the normal matter protons. They automatically circle clockwise in the machine while the protons circle counterclockwise. It is relatively easy to keep the beams separated a few centimeters from each other while keeping them in the vacuum pipe, and then bringing them together at the desired collision stations where the detectors are waiting to detect the resultant particles. In this scheme, antiprotons are no longer the goals of the research, but merely engineering tools being used by the scientists to attain the next goals in particle physics.

This cheaper technique was the route taken by the European scientists in the 1980s to find the predicted W and Z particles that are the carrier particles for the weak nuclear force. Led by Simon van der Meer and Carlo Rubbia, the European scientists snatched the Nobel Prize away from the American scientists before they could get their proton-proton machine operating. In the process, the Europeans developed all the tools needed to make the mass production of antiprotons and antimatter a firmly based technology instead of an inexact science.

Making and Storing Antiprotons

The basic method of making antiprotons is shown in Figure 5–1. Normal matter protons are accelerated to high velocities in a proton synchrotron accelerator. Their kinetic energy (that is, energy of motion) quickly becomes many times greater than their rest mass energy (a little less than 1 GeV). Scientists at CERN in Switzerland accelerate protons until their kinetic energy is 26 GeV, some 26 times greater than their rest mass. At the Institute for High Energy Physics (IHEP) in the Soviet Union, the protons reach a kinetic energy level of 70 GeV. At Fermilab in the United States, the proton

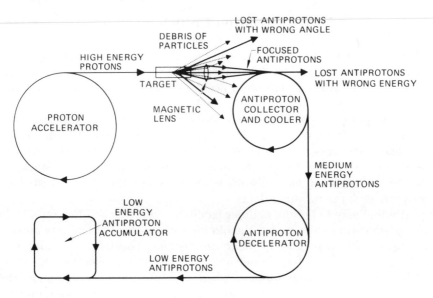

Figure 5-1 The present method for making antiprotons.

synchrotron accelerator is operated at 120 GeV. A bullet proton there has over 120 times as much kinetic energy as rest mass energy. That means a high-speed proton has enough kinetic energy in it to make 120 copies of itself, or 60 pairs of protons and antiprotons.

The high-energy bullet protons then slam into a metal target. They collide head-on into the metal nuclei. Their kinetic energy is released at a single point in space in a tiny fireball. The kinetic energy turns into a spray of photons (namely, X rays and gamma rays) and pairs of particles and antiparticles. A small fraction of those pairs are protons and antiprotons. At CERN, with its 26 GeV machine, only 0.4 percent (four tenths of one percent) of the bullet protons manage to produce a proton-antiproton pair. At IHEP the production efficiency is about 2 percent. At Fermilab the efficiency is 4.7 percent, nearly 12 times greater than CERN's.

Just making antiprotons is not enough. The researchers must next capture them and put them in a bottle. It isn't easy. The antiprotons come out of the target in a real spray. The bullet protons collide with metal nuclei and create antiprotons at many different points in the copper wire target. Most of the antiprotons are emitted traveling in nearly the same direction as the bullet protons. However, a good-sized fraction of the antiprotons come out of the target at a wide range of angles. They also come out of the target with a wide range of energy levels. At Fermilab the antiprotons have an energy that peaks around 12 GeV, but the spread is from 3 to 50 GeV. This spray of antiprotons—in energy, angle, and point of origin—has to be separated from similar sprays of electrons and positrons, pions and antipions, neutrons and antineutrons, and other pairs of particles and antiparticles. Fortunately, the antiprotons have a mass and an electrical charge that is unique, making it possible to identify them and separate them from the rest of the particles created in the collision of the bullet protons with the target.

To collect the antiprotons coming out at different angles, researchers use a *magnetic lens*, an arrangement of magnetic fields that curves the paths of the moving antiprotons. In this way the diverging spray of antiprotons is focused again on the other side of the lens. Magnetic lenses are not terribly effective. Antiprotons that leave the target at a wide angle are much less likely to get focused back into a line than are those that came out of the target at a narrow angle. Usually only 10 to 30 percent of the antiprotons created in the target (which itself is not more than 5 percent efficient!) are captured by the magnetic lens and brought to a focus.

The next step is to collect the antiprotons into a "bottle" that can hold them. The bottle currently used at particle accelerators does not look very bottle-like. It is actually a long pipe with a vacuum inside it, bent into the shape of a ring. Bending magnets surrounding this collecting ring bend the paths of the antiprotons moving through the vacuum pipe. If the speed and direction of the antiproton, the strength of the magnetic field, and the radius of the bend of the vacuum pipe are just right, then the path of the antiproton will be bent just the right amount to travel down the center of the vacuum pipe without hitting the walls. A typical vacuum pipe is about 10 cm (4 inches) in diameter and hundreds of meters long. Threading the beam of antiprotons through it is quite difficult. The best capture efficiency anyone has been able to accomplish to date is only one percent. Ninety-nine percent of the antiprotons, with the wrong energy, end up in what physicists call the "beam dump"—they are lost forever.

When the antiprotons have been captured in the collecting ring, most of the hard work has been done. The vacuum level in collecting rings is so low that the antiprotons can be stored for days at a time. Scientists can add more high-energy antiprotons at any time and increase their number. A typical antiproton collector ring holds as many as a trillion (10^{12}, or 1,000,000,000,000) antiprotons at any one time. A single antiproton has a mass of about 1.7×10^{-24} grams (that's 5.9×10^{-26} ounces, or about 600 millionths of a billionth of a billionth of an ounce). So a million million of them has a mass of about 1.7×10^{-12} grams, or 1.7 picograms. This may not sound like much antimatter. But if 1.7 picograms of mirror matter were annihilated with an equal amount of matter, the total energy released would be about 300 joules. It's enough energy to light a 100-watt lightbulb for three seconds. It's enough energy for you to see.

Scientists are always looking for ways to improve the storage performance of an antiproton collector. They also look for ways to make it easier to use the stored antiprotons. For these reasons, the collector ring usually has devices that "cool" the stored antiprotons. When the antiprotons are first made, they have a wide spread of energies and velocities. Some antiprotons are moving faster, and some slower, than the average velocity. Particle physicists use various techniques to get all the antiprotons to travel at the same velocity in the collector ring. They still have lots of kinetic energy because of their high average speed, but the "temperature" is low because the variations in their energy are small. They are "cooler" than before.

Now that the antiprotons are cooled, they can be send to an antiproton decelerator, which is really nothing more than a proton accelerator run backwards. At CERN, the proton accelerator first speeds protons up to about 26 GeV to make antiprotons. The antiprotons are captured in a collector ring. It takes about two seconds to cool down the pulse of antiprotons. While that's going on, the proton accelerator carries out a few tasks, and then is reconfigured as an antiproton decelerator. Now it slows the cooled 3 GeV antiprotons down to about 500 MeV. The antiprotons can then be captured and stored in a machine called a low-energy antiproton accumulator. Once the engineers have all the machines tuned up correctly, the efficiency of cooling, transferring, decelerating, and storing the antiprotons at low energy is quite high—more than 90 percent, in fact. But considering the terrible losses during the initial processes of making and capturing antiprotons, the overall losses are really quite high.

For a typical "factory," of all the bullet protons that enter the target, only about one percent make antiprotons. Of those antiprotons, only about ten percent come out at the proper angles to be focused by the magnetic lens. And of those focused antiprotons, only another one percent are captured in the antiproton collector ring. So the typical antiproton production facility has a total efficiency of one antiproton for every 100,000 high-energy bullet protons. There is one additional factor to keep in mind: The typical proton synchrotron accelerator is itself less than ten percent efficient at turning electricity into proton beam energy. So the overall efficiency of of the whole process is less than one part in a million.

Several things can be done to improve by several orders of magnitude the present abysmal efficiency of antiproton production. Before we look at those methods, though, let's examine the ways various facilities around the world today accomplish the tasks needed to make, collect, and store antiprotons.

Antiproton Physics at CERN

The oldest major producer of antiprotons in the world today is the European Organization for Nuclear Research, located outside Geneva, Switzerland. With France as the major partner, its original name was the Centre Européenne pour la Recherche Nucléaire, or CERN. The acronym has stuck.

CERN runs three large machines used in elementary particle physics: the Proton Synchrotron (PS), the Intersecting Storage Rings (ISR), and the Super Proton Synchrotron (SPS) (Figure 5–2). The PS is a proton accelerator that boosts protons up to 26 GeV. The ISR consists of two rings that store protons accelerated by the PS. The two stored beams travel in opposite directions and intersect at four different points. Protons collide

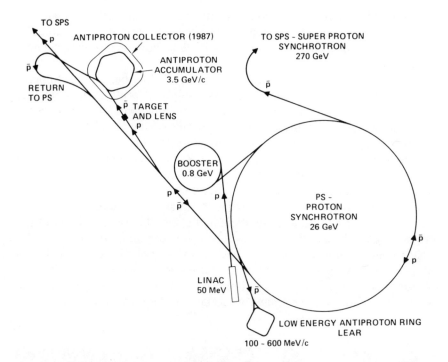

Figure 5-2 The facilities for making and storing antiprotons at CERN are complex, with various booster, cooling, and storage rings. European physicist Carlo Rubbia won the Nobel Prize in Physics for his engineering improvements to CERN, which resulted in the discovery of two new subatomic particles.

with protons at those points, and the CERN researchers analyze the debris from the collisions using complex detectors located at the collision points. Finally, the SPS takes 26 GeV protons from the old Proton Synchrotron and boosts them up to 270 GeV. The SPS is more than 2 kilometers (1.4 miles) in diameter, or nearly 7 kilometers (4 miles) in circumference. The CERN complex will grow with the future addition of a *Large Electron-Positron* (LEP) collider ring, which will be 28 kilometers (17 miles) around.

The process of making antiprotons at CERN begins with the gathering of some protons. An electric discharge strips the electrons from atoms of hydrogen, leaving the proton nuclei behind. The protons are extracted from the hydrogen discharge chamber and injected into the linear accelerator, or LINAC. The LINAC accelerates the protons up to an energy of 50 MeV and injects them into the Booster. The Booster then boosts them up to a near-relativistic energy of 800 MeV. That makes their kinetic energy comparable to their rest mass energy of 938 MeV. The Booster then passes the protons on to the Proton Synchrotron, the heart of the antiproton production process at CERN. The PS boosts the protons to an energy of 26 GeV. Now they can be sent to the ISR for storage and proton-proton collisions. They can also be injected into the SPS for further increases in energy to 270 GeV, and more experiments. Or, the protons can be switched to the target area to make antiprotons (Figure 5–3 shows a proton accelerator at another facility.)

The targets at CERN are short wires of metals like beryllium, copper, or tungsten, about 10 centimeters (4 inches) long by 2.5 millimeters (0.1 inch) in diameter. The protons are sent streaking down the wire in a narrow beam to interact with the metallic nuclei. The protons scatter the electrons around the nuclei and deposit a large amount of energy in the small diameter wires. The target wires are embedded in a heat sink with high thermal conductivity cooled by large flows of cooling water. Even so, some of the early targets melted or even exploded under the impact of the pulsed proton beams.

As the rapidly moving protons strike the heavy metal nuclei, their kinetic energy is converted into a spray of elementary particles, including antimatter particles such as antiprotons, antielectrons, and antineutrons, moving at speeds close to that of light. You would think that the task of capturing the rapidly moving antiprotons from the cloud of debris emanating from the tungsten target would be as impossible as trying to catch the queen bee in the swarm of bees emanating from a kicked-over hive. But with the

Figure 5–3 The interiors of all giant proton accelerators look much the same. This is the interior of the Alternating Gradient Synchrotron (AGS) at the Brookhaven National Laboratory, which accelerates protons to within one percent of the speed of light. This photo shows a portion of the AGS ring. Leon Lederman used the AGS to discover the antideuteron in 1965. The muon and tau neutrinos were also discovered with the AGS. (*Courtesy Brookhaven National Laboratory*)

aid of a lens made of magnetic fields and a magnetic particle selector, the negatively charged antiprotons can be separated from the remainder of the debris consisting of particles that have a different charge and mass than the antiproton.

At CERN the magnetic lenses that have been used to focus the spray of antiprotons to a point where they can be collected come in two types. One is called a *magnetic horn* (Figure 5-4) and the other is the *lithium lens* (Figure 5-5). The magnetic horn is a long cylinder of copper shaped like two ancient herald-type straight horns joined at the mouthpieces. A very powerful current enters through one bell, passes through the narrow "mouthpiece" junction, and exits through the other bell. The strong current passing through the horns creates a strong magnetic field outside the horn. Because of the symmetry of the lens and Maxwell's equations for magnetism, there is no magnetic field inside the horns. Those antiprotons exiting from the target that are going straight ahead, pass right along the axis of the magnetic horn, and pass through the magnetic lens without encountering any magnetic field. Those antiprotons that leave the target in a direction that diverges from the axis sooner or later pass through the first thin copper horn and encounter the strong magnetic field outside the horn. The magnetic field bends the antiprotons back toward the axis, where they pass back through the metal in the second horn. They meet the on-axis antiprotons at the focal point. Those antiprotons that emerge from the target at still higher angles off the axis, pass through the first horn sooner, encounter the magnetic field for a longer time, and get bent back proportionally more, so they too reach the focus along with all the other antiprotons. Thus, by proper shaping of the diameter and thickness of the copper in the bells of the horns, it is possible to take a spray of antiprotons coming from some small target region and

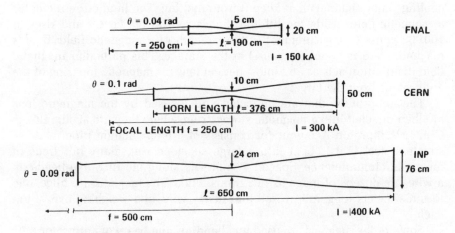

Figure 5-4   A parabolic horn magnetic lens is used by physicists to focus the spray of antiprotons to a point where they can be stored.

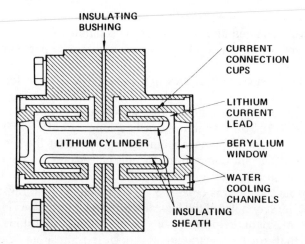

INSULATING
BUSHING

CURRENT
CONNECTION
CUPS

LITHIUM
CURRENT
LEAD

LITHIUM CYLINDER

BERYLLIUM
WINDOW

WATER
COOLING
CHANNELS

INSULATING
SHEATH

Figure 5-5   The lithium magnetic lens, invented by Russian physicists, is even better at focusing antiprotons than the magnetic lens.

focus them back down to a point where they can be inserted into a storage ring.

Tanya Vsevoloskaya, one of the best particle accelerator engineers in the U.S.S.R. (if not the world), invented an even better magnetic lens called a *lithium lens*, now being used at CERN and Fermilab. It consists of a cylinder of lithium, a very light metal—the one at CERN is a cylinder about 5 millimeters (1/4 inch) in diameter and about 10 centimeters (4 inches) long. A huge current pulse, many thousands of amperes, is sent along the axis of the lens in coincidence with the antiproton pulse. The lithium is heated almost to its melting point by the pulse, but there are high-pressure cooling water channels that keep it from melting. The high current creates a magnetic field inside the lithium that is zero at the center and rises to 100,000 *gauss* (a measurement of the strength of a magnetic field; Earth's magnetic field is about 0.5 gauss) at the surface. This particular magnetic field distribution acts as an almost perfect lens for magnetic focusing of the antiprotons from the target.

The antiprotons that are collected and focused by the magnetic lens are then directed into a magnetic storage ring, where they are accumulated. CERN's Antiproton Accumulator consists of a long vacuum pipe made of stainless steel that is bent into a large six-sided ring many hundreds of meters in circumference. Spaced at intervals along the ring, like beads on a wire, are dozens of magnets of various kinds. The magnets are huge, one to three meters long, full of iron and heavy wire, and weighing over a ton each.

Some of the magnets are used for "bending" the beam of antiprotons. As the negatively charged antiproton passes through the magnetic field penetrating the vacuum pipe from top to bottom, the charge on the antiproton experiences a sidewards force that is at right angles to both the direction of the magnetic field and the direction of travel of the antiproton. This sidewards force causes the path of the antiproton to be bent slightly so the antiproton follows the curvature of the vacuum pipe. At CERN, the bending magnets are concentrated at six places along the Antiproton Accumulator ring, giving it a rounded hexagonal shape. The straight sections of the Accumulator are where the antiprotons are "cooled" (more about that in a moment).

Other magnets in the storage ring use more complex magnetic field shapes to keep the beam focused in the center of the vacuum pipe so none of the antiprotons strike the walls and are lost. There are also smaller, stronger magnets that can be pulsed to move the beam around inside the vacuum pipe, or to switch the beam out of the ring and into an extraction pipe when it is time to do experiments with the antiprotons. If everything is lined up properly, the antiprotons circle endlessly around and around inside the Accumulator ring. The scientists at CERN in Switzerland have stored antiprotons in this manner for days.

The antiprotons enter the Accumulator every 2.4 seconds in bursts of about a trillion (1,000,000,000,000) at a time. Only about five million make it all the way around the Accumulator ring without hitting the walls. Each burst of antiprotons is "stacked" near the inner wall of the ring before the next burst arrives. The stacks slowly build up to a core of about a hundred billion antiprotons. About a trillion antiprotons can be stored in the Accumulator using current cooling techniques.

Cooling Antiprotons

The antiprotons generated in the tungsten target have a wide spread of energies. Before they can be used further, it is necessary to "cool" the antiproton beam so that all the antiprotons have the same velocity. CERN, of course, uses the cooling technique invented by CERN's own Carlo Rubbia, called *stochastic cooling*. *Stochastic* is a scientific word meaning "random" (even though the Greek root word means "skillful in aiming"!). As the antiprotons move around inside the vacuum pipe at different velocities, they will pass each other and "clumps" of charged antiprotons will temporarily form. These clumps of charge are detected by a current sensor. The signal from the sensor is amplified, switched in sign, then quickly sent by cable from one side of the ring to the other. Since the cable cuts across the center of the ring, while the antiprotons have to circle around the circumference of the ring, the signal arrives before the clump of antiprotons, even though the antiprotons are moving at nearly the speed of light. The signal then drives a "kicker" that uses an electrical pulse to accelerate or decelerate the "clump" of current, smoothing it out. By using hundreds of sensors and kickers, the scientists and engineers at CERN can decrease the randomness in the antiproton energies by an *order of magnitude* (ten times) in less than two seconds. They also use other sensors to detect antiprotons which have the correct average speed but are not moving straight down the vacuum pipe, instead snaking from side to side. Signals from those sensors are then sent to transverse kickers, which straighten out the antiproton paths.

After this preliminary cooling process using stochastic cooling, the antiprotons are switched out of the capture and stochastic cooling ring into a particle decelerator. At CERN, the antiproton decelerator is just the PS run backwards. The antiprotons enter the decelerator with an energy of 3000 MeV and leave the decelerator with an energy of 200 MeV. This fifteen-fold decrease in energy means that the antiprotons, which used to be moving at relativistic velocities, are now moving at speeds well below the speed of light.

The decelerated antiprotons then go to another storage ring called the Low Energy Antiproton Ring (LEAR)—actually not a ring but a square, with sides of about 20 meters and rounded corners. At each corner is a set of bending magnets that take the antiproton beam, turn it through an angle of 90 degrees, and send it shooting down the next straight section of vacuum pipe to the bending magnets at the next corner.

The antiprotons have a low spread of energies after they are stochastically cooled in the antiproton accumulator. After stochastic cooling, but before deceleration, the spread in energy is reduced to 0.2 percent of the average energy of 3000 MeV, or 6 MeV. After the antiprotons are decelerated to 200 MeV, however, that 6 MeV spread is 3 percent of the average antiproton energy. Thus, after deceleration, it is necessary to cool the beam of antiprotons further. CERN scientists use stochastic cooling in the LEAR machine for further cooling

The cold beam of low-energy antiprotons is then used for various scientific experiments, such as measuring the production rates of some of the more exotic particles created by proton-antiproton annihilation. The frequency with which rare and exotic particles are created tells scientists a great deal about the inner quark structure of protons and antiprotons.

Other experimenters use the low-energy antiprotons from the LEAR machine to study the X rays created just before annihilation takes place. When the antiprotons in the LEAR machine penetrate the target in the center of the X ray experiment detector, they hit electrons in the target and slow down. As they do so they lose energy. When the antiprotons have nearly come to a stop, they are moving slowly enough that they "sense" the presence of the positively charged protons in the nuclei of the target atoms. The antiproton is negatively charged, and so it acts like a "heavy electron" (1,836 times as heavy as a real electron). It begins to orbit the positively charged nucleus, starting in a high-energy orbit and jumping down from there into lower-energy orbits. Each time it jumps it emits X rays. Finally, the antiproton is close enough to the nucleus that annihilation takes place.

Researchers have learned interesting things by studying the release of these X rays. For example, the pattern of X rays emitted before annihilation is different for each element. This could have practical significance for analyzing the composition of unknown materials and alloys, as we'll see in Chapter 8.

The LEAR machine typically holds about 10 billion antiprotons, enough to supply the needs of half a dozen experiments for many hours. Of course, not all the antiprotons get used. About ten percent of them (a billion or so) end up in the LEAR's "beam dump." With a little planning, the beam dump could be replaced with an antiproton trap. Antiprotons previously lost forever could be saved and used for other experiments. The first step towards this has already been taken.

Storing Antiprotons

One of the first questions people have when first introduced to the concept of antimatter is, "How do you 'hold' onto this magical mirror matter that disappears in a burst of energy the instant it touches normal matter?"

The scientists in Switzerland have demonstrated one solution to this problem. Their "bottle" is an evacuated tube about 5 centimeters (2 inches) in diameter and bent into a ring about 300 meters (984 feet) in circumference. As the beam of antiprotons goes around, it is directed and focused by magnetic fields to keep it from hitting the walls of the tube.

Scientists working in atomic physics and plasma physics have come up with a much cheaper, lighter, and smaller trap for mirror matter, called a Penning trap (Figure 5-6). The side walls of the Penning trap consist of a carefully machined solid metal ring about 5 centimeters (2 inches) in diameter. The inside has a hyperbolic shape. Above and below the hole in the ring are two domed metal end caps, also with hyperbolic shapes. This trap is placed in a vacuum chamber inside the bore of a superconducting magnet in a thermos jug containing liquid helium. The magnetic field from the superconducting magnet runs along the axis of the trap from one end cap to the other.

To capture an antiproton once it has been inserted into the trap, the end caps are given a negative charge and the ring a positive charge. The negative charge on the caps repel the negatively charged antiproton, keeping it from going in the axial direction. The antiproton will attempt to move radially outward toward the positively charged ring electrode, but the magnetic field will keep it moving in a circle. If the magnetic field is strong enough and the trap is cold enough, the antiproton will never get to the ring and it will be permanently trapped in a tiny circular orbit.

These compact traps have already demonstrated their ability to store mirror matter. At the University of Washington in Seattle, one of these traps held a positron for over a month, demonstrating that the trap is stable over long periods of time and the vacuum inside a supercold container is extremely good. Gerald Gabrielse, now at Harvard University, recently demonstrated that these Penning traps can also hold antiprotons.

For years, Gabrielse has been storing individual protons and electrons

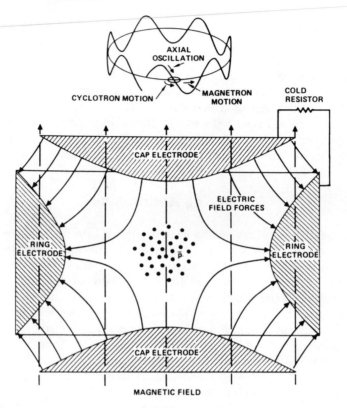

Figure 5-6   A Penning trap is a device that can trap and hold antiprotons using a combination of steady electrical and magnetic fields.

in carefully designed Penning traps and measuring their fundamental properties. He holds the particle in the trap and deliberately excites it into a tiny vibration, then measures the frequency of the vibration. By knowing the "spring" characteristics of his trap and the frequency of excitation, he is able to measure the mass, charge, or magnetic moment of the particle to extreme precision.

In July 1986 Gabrielse led a group of students and colleagues to the Low Energy Antiproton Ring (LEAR) at CERN, where they finally succeeded in trapping and holding antiprotons in a Penning trap. CERN had given Gabrielse's group only one day on the LEAR machine to do a feasibility test for some precision antiproton measurements they planned to carry out in the near future.

Using a relatively crude Penning trap with a thick foil degrader at the entrance, they slowed the 21 MeV antiprotons from the LEAR machine more than four orders of magnitude down to 1 keV. The antiprotons were captured in flight by rapidly applying kilovolt potentials to previously grounded trap electrodes while the antiprotons passed through the trap. During the test period, they repeatedly demonstrated the trapping of as many as 200 antiprotons out of single 150–nanosecond pulses of antiprotons sent to the trap from LEAR. In a typical run, the antiprotons were held 100 seconds, then allowed to escape the trapping region to annihilate on a wall so the time-of-flight spectrum of the detected pions from the annihilation of the antiprotons could be measured to prove that there had been antiprotons in the trap.

The experiment was completed successfully despite the fact that they were working against time (only 24 hours on the LEAR machine) and adversity (something went wrong with their trap and it had to be cycled up to room temperature just before their 24 hours was to commence); the observed holding times were greater than 100 seconds—in one case, 10 minutes. These times are already long enough to permit holding the kilovolt antiprotons until the effect of the high-intensity pulse has decayed away, then sending the trapped antiprotons into a high-quality storage trap or an interaction region.

More experiments are planned for the near future, after the resumption of operation of the improved LEAR facility. The purpose of these experiments is not to demonstrate the trapping of mirror matter—in fact, if the scientists accidentally trap more than one antiproton, they will eject all but one. The purpose is to measure the mass of the antiproton to one part in 100,000,000,000, and the experiment is more accurate if there is only one antiproton in the trap.

Everyone knows that the antiproton mass should be exactly the same as the normal proton mass; a finding that it isn't would be very significant. Future experiments might include the direct measurement of the gravitational acceleration of an antiproton or the synthesis of antihydrogen.

There is a more ambitious program under way to capture an even larger number of antiprotons in a Penning trap. The ultimate purpose of the experiment is to measure the effect of gravity on the antiproton to see if it falls faster than, slower than, or at the same speed as a proton. The experiment consists of obtaining antiprotons of the lowest energy possible from the LEAR facility at CERN, decelerating them further in the external beam line using a special antiproton decelerator called a radio frequency quadrupole, trapping them in a large Penning trap (with cylindrical walls, since precision isn't important), cooling them to ultralow energy using a number of electronic cooling techniques, then measuring their gravitational acceleration by "tossing" them up a vacuum tube and timing their trajectory.

The experiment has been granted approval by the CERN experiment review board and the apparatus is presently under construction. The experimental team consists of 32 collaborators from the Los Alamos National Laboratory, several universities, and the University of Pisa in Italy. These last were important for at least historical reasons, for it was at the Tower of Pisa that Galileo was supposed to have carried out his famous experiment of dropping balls of different materials to prove that gravity acted equally on all objects. The experimenters will not be able to drop a proton and an antiproton off the Leaning Tower of Pisa; they will have to carry out their experiments in a long vertical vacuum pipe in nearby Geneva, Switzerland.

The team has constructed a test beam line at the Ion Beam Facility at the Los Alamos National Laboratory in New Mexico to begin developing the apparatus for the experiment. To test the apparatus they use a negative hydrogen ion, which is a hydrogen atom with an extra electron, or a single proton with two electrons in orbit arount it. The negative hydrogen ion has the same charge as the antiproton and almost the same mass, so it makes an excellent test particle. A negative hydrogen ion beam is obtained from an existing source and sent through a horizontal section, then turned in a vertical direction by a magnet. The beam will then be steered and focused into a Penning trap situated in the superpowerful field of a superconducting solenoid magnet.

The planned demonstration technique is to run the beam into the trap with enough voltage on the upper end cap of the trap to repel the beam, and then to pulse the lower cap to sufficient voltage to trap the ions. Several hundred milliseconds later the voltage on the upper cap will be dropped to release the trapped ions upward to be detected in a microchannel plate at the top of the short drift tube. The time of flight to the top of the drift tube will allow the gravity acceleration on the antiprotons to be compared to the gravity acceleration on the similar normal matter negative hydrogen ions.

It was originally thought that these electromagnetic Penning traps could only hold charged particles at extremely low density. It would take a trap as big as a walk-in refrigerator to hold a microgram of antiprotons. A group of US plasma physicists, however, have found that under certain conditions, these kinds of cold magnetic traps may allow a rotating gas cloud of charged particles to condense into a rotating "liquid" ball of charged particles or even a rotating dense "solid" of charged particles. Many micrograms of mirror matter could then be stored in small "fuel tanks" no bigger than a thermos jug.

Antiproton Physics at IHEP

High-energy particle physics in the USSR is carried out at the Institute for High Energy Physics (IHEP) in Serpukhov, Novosibirsk (Figure 5–7). The major machine is the U–70 proton synchrotron, with a maximum energy of 70 GeV and a beam intensity of about 7×10^{12} (7,000,000,000,000) protons per cycle. Each accelerating cycle takes about seven seconds to bring the protons up to the maximum energy. The IHEP researchers plan to build a Very Large Electron-Positron Collider machine with an energy of about 1,000 GeV, or one TeV (one tera-electron-volt). By 1990 they also plan to have one ring of an Accelerating and Storage Complex (ASC) operating, with protons and antiprotons colliding with a maximum energy of 6 TeV. The ASC will be built in an underground tunnel, like the accelerators at

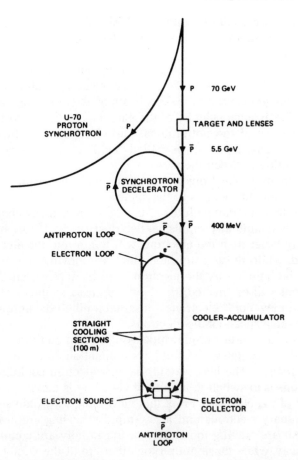

Figure 5-7 The arrangement of machines used to make antiprotons at the IHEP Laboratory in the Soviet Union.

CERN, Fermilab, and other locations around the world. The tunnel for the Russian ASC will be 19 km (12 miles) in circumferance.

IHEP will use the 70 GeV protons from the U–70 machine to make antiprotons. The proton beam is focused down to 0.5 millimeters (0.02 inches) with a lithium lens. The beam strikes a target made of metal wire. Soviet scientists have experimented with many kinds of target material, including beryllium, copper, and tungsten wires. They have also been working with a target that is actually a continually flowing stream of mercury. The flowing mercury target can handle higher intensity proton beams, because it continually supplies cool fresh target material.

The antiprotons generated in the target have a peak energy of about 5.5 GeV. They are collected by a second lithium lens and then go on to a synchrotron-decelerator. There they are slowed from 5.5 GeV to 400 MeV of energy. Then the antiprotons are sent to a cooler-accumulator to be cooled and stored. The IHEP cooler-accumulator looks like a racetrack. The ends are two half-rings with a radius of 40 m (130 feet) and two straight sections 100 m (328 feet) long.

Soviet scientists at IHEP are using a cooling technique they themselves invented called *electron cooling*. To carry out electron cooling, a carefully designed electron gun is used to make a beam of electrons that all have the same energy. The average velocity of the electrons is chosen to be exactly the average velocity of the antiprotons. The electrons vary little around their average velocity; they can be called "cold" electrons since there is little relative motion between them.

This cold stream of electrons travels along with the beam of "hot" antiprotons in the straight sections of the cooler-accumulator. The negatively charged antiprotons interact with the much lighter negatively charged electrons through their mutually repulsive electrical fields. Those antiprotons that are moving faster than the electrons will bump into the electrons and be decelerated, while those antiprotons that are moving slower than the electrons will be bumped by the electrons and be accelerated. The much smaller electrons will be "heated" during the process, so the electron beam is removed and new "cold" electrons reinserted until all the antiprotons are moving at the same speed as the electrons.

The electron cooling technique works equally well on those antiprotons that have the right average speed, but have a direction that is at an angle to the average direction. The best way to see how electron cooling works on these antiprotons is to switch to a point of view that is moving along at the average speed of the electrons. From that viewpoint, we would see a cloud of nearly stationary electrons and a few rapidly moving antiprotons. The antiprotons that are moving too slowly or too rapidly are coming toward us or going away, while those antiprotons that are at the wrong angle are weaving from side to side through the nearly stationary cloud of electrons. The moving antiprotons bump into the electrons and "heat" them up, but are themselves "cooled" in the process. At the ends of the cooling track the antiprotons, with their greater momentum, leave the ring. They are turned around by the antiproton bending magnet channel and are sent back along the other straight section for further cooling.

The IHEP researchers have plans to improve the production of antiprotons at their facility. One concept under study would increase the intensity of the proton beam by first stacking up seven bunches of protons. The bunches would then all hit the target at once. Provided the target doesn't explode, a single pulse would generate seven times as many antiprotons. A second concept would involve decreasing the velocity spread of the protons by stochastic cooling or some other means. That would allow four proton bunches to be stored where previously only one could. A third concept would use a multiple target system. The proton beam would pass through several short targets in succession. Lithium lenses would separate the targets from one another. Some protons would generate antiprotons in the first target, some of the remainder would create antiprotons in the second, and so on, until all the protons are used up. The lithium lenses would focus the antiprotons generated. Most would pass through the remaining targets to be focused finally into one beam of antiprotons.

If all these improvements are carried out, IHEP could improve its rate of antiproton production 140 times over its current rate. In fact, if this improved IHEP antiproton production "factory" ran 24 hours a day, the Russians could produce 0.3 micrograms of antiprotons per year. The energy of 0.3 micrograms of antiprotons annihilating with an equal amount of protons is 50 million joules (megajoules). That's enough energy to run a rocket engine for several minutes.

Antiprotons at Fermilab

Many particle accelerators in the United States can be used for physics experiments. The most powerful accelerators, though, are at the Fermi National Accelerator Laboratory (Fermilab), about 40 miles southwest of Chicago (Figure 5–8). Fermilab's Main Ring, its older high-energy proton accelerator, is capable of boosting protons to 400 GeV. U.S. scientists and engineers have now completed a new machine capable of taking the 400

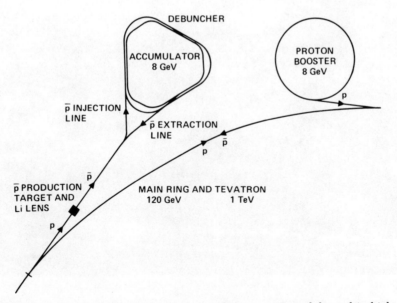

Figure 5-8   The arrangement of particle accelerators at Fermilab used in high-energy particle physics. Fermilab's proton collider machines make antiprotons at energy levels between 10 and 400 GeV. The Tevatron machine planned at Fermilab will have energies higher than 1,000 GeV.

GeV protons and accelerating them up to 1,000 GeV, or 1 TeV—a tera-electron-volt. The new machine, not surprisingly, is called the Tevatron.

The Tevatron has superconducting magnets. They not only use much less electrical power, but also produce a much stronger magnetic field than can standard magnets. The stronger magnetic field can bend the paths of the protons more. That means the Tevatron can use a tunnel the same size and radius as the old Main Ring tunnel. In fact, the Tevatron is tucked underneath the old Main Ring. Figure 5–9 shows the Tevatron on the floor of the tunnel below the Main Ring.

American scientists had long intended to use the Main Ring and the Tevatron to find new high-energy subatomic particles. The Main Ring would supply the Tevatron with protons, and the Tevatron would boost the protons to 1 TeV. Then the Main Ring would accelerate a second batch of protons to 400 MeV. Scientists would collide the 400 MeV protons with the 1 TeV protons. They were looking for particles called the W^+, the W^-, and the Z^0. The unified field theory of Sheldon Glashow, Steven Weinberg, and Abus Salaam predicted their existence.

But before the Americans could get the Tevatron built and running, the Europeans at CERN made an end run around them. Using their 270 GeV SPS machine, they found the predicted particles first. Two members of the

Figure 5-9 A portion of the accelerator tunnel at Fermilab. The magnets in the upper ring are for the 400 GeV Main Ring accelerator. The superconducting magnets in the lower ring are for the 1,000 GeV Tevatron accelerator. (*Courtesy Fermi National Accelerator Laboratory*)

European team, Carlo Rubbia and Simon van der Meer, won the 1984 Nobel Prize in Physics for their brilliant work. The SPS used much less energy than the Tevatron. The Europeans managed to pull off the end run by using a secret weapon.

Antimatter.

Instead of building their own Tevatron, Rubbia and van der Meer and their confreres found it was faster and cheaper to build an antiproton production facility. They made antiprotons to use in the 270 GeV machine and turned it into a proton-antiproton collider. The resulting machine had barely enough energy to create W and Z particles. But it *was* enough. Rubbia and van der Meer confirmed the Glashow-Weinberg-Salaam electroweak unified field theory. And they got the Nobel Prize.

After that shocking setback, the Americans at Fermilab started building their own antiproton facility. That would make it possible to turn the Tevatron into a proton-antiproton collider that could reach 2 TeV of energy. Perhaps the Fermilab scientists could then find their own new antiparticle. And if there were a Nobel Prize in Physics for them, so much the better.

The antiproton facility at Fermilab is pretty simple. Protons with an energy of 8 GeV from the Proton Booster are sent to the Main Ring. There they are accelerated up to 120 GeV. (The Main Ring can kick protons up to about 400 GeV, but for a number of highly technical and somewhat obscure reasons it's best for antiproton production to go only to 120 GeV.) The 120 GeV protons then smash into a metallic target and produce antiprotons. At Fermilab the target is a rotating cylinder of tungsten metal. The antiprotons from the target are collected by a lithium lens, and a pulsed magnet separates the focused antiproton bunches from the residual protons and other subatomic particles coming out through the lens. The antiprotons get sent to a "Debuncher ring," really a triangle with rounded vertices. Here the bunches of antiprotons are "cooled" down. Electromagnetic fields take a bunch of antiprotons with a narrow pulse length and large velocity, and "stretch" them out. That cools them down by decreasing their spread in velocity. After the debunching/cooling process, the antiprotons are stochastically cooled even further, and passed on to the Accumulator. Here successive batches of antiprotons are stacked. After about four hours the Accumulator contains up to 400 billion antiprotons.

When the Accumulator is full, antiproton bunches can be extracted, sent back to the Main Ring, accelerated up to 150 GeV, and injected into the Tevatron. There they are smashed head-on into a beam of protons. Out of those annihilation collisions may come previously unseen subatomic particles.

Future plans at Fermilab call for adding more precooling to the Debuncher; improving the stochastic cooling in the Accumulator; adding some electron cooling devices to the Accumulator; and improving methods for extracting particles from the Main Ring for antiproton production. There are, however, no plans to build a low energy antiproton ring at Fermilab (like the LEAR ring at CERN), or to modify the Fermilab machines in such a way that antiprotons could be slowed down and captured.

The Rush to Win

The production efficiency of the present machines are abysmally low. They range from parts in a billion in Switzerland to parts in a million in the U.S.S.R. Fortunately, the low efficiency is not due to any fundamental limitation, and can be improved by orders of magnitude. The reason for the low efficiencies is that all of the mirror matter production facilities to date have been built as crash projects under limited budgets. Production efficiency has been sacrificed to such considerations as speed, cost, science requirements, and national pride.

Because of the desire to be first to reach new energy levels, and the limited budgets, many shortcuts have been taken in the designs of the mirror matter factories. For example, in Switzerland, scientists had to use the proton accelerator machine they had available even though they knew it generated protons with marginal energy for producing antiprotons. If they had had a machine of the right energy, they could have produced 20 times the number of antiprotons per proton. Then, of the antiprotons coming out of the target, they only manage to collect one out of a thousand in the collector ring. The rest are allowed to get away because the scientists could only afford one focusing lens and one capture ring. An array of magnetic lenses would allow the capture of antiprotons coming from the target area at many different angles, while a typical ring tunnel could hold many capture rings, each designed to capture antiprotons over a different range of energies.

Another factor that is limiting the research groups is that their proton accelerators are designed for precision, not production. The proton current in these machines is designed to be low so that the machines can produce protons of extremely high and quite precise energy. Thus, when the protons strike a target, the scientists know their exact energy and can make precise measurements on the new particles that are produced so they can compare the measurements with the predictions from the latest version of elementary particle theory. As a result, these research tools have voltages that are *too high* and proton currents that are *too low* for optimum antiproton production. The machines are also not very efficient; typically, only five percent of the electrical energy from the power lines ends up as energy in the proton beam.

There are designs for proton accelerators that are much more energy efficient, called *linear accelerators*. As their name implies, instead of being built in a large circle, they are built in a straight line. In the circular proton accelerator rings, the protons circle around and around the ring, each time getting a kick from the single section of radio energy kickers. In a linear accelerator, there are many more radio energy accelerating sections and the protons only pass through once. Such linear accelerators would have an energy efficiency of 50 percent from the "wallplug" to energy in the proton beam. You can't do much better! A linear accelerator needed to produce protons with the optimum energy to make antiprotons, however, would be quite long—40 kilometers (25 miles). But there is plenty of room in the deserts of Texas and New Mexico, and there is even more room in space.

A 40 kilometer-long antiproton factory would be a little longer than the 28-kilometer circumference LEP ring presently under construction at CERN and one fifth the size of the 200-kilometer circumference supercollider being proposed as the next large particle accelerator in the United States. If run at a power level of 10 GW, the proton current required would be only 1/20 of an ampere. By separating out the antiprotons coming from the target and dumping the remaining particles into an electronuclear breeder reactor loaded with unenriched uranium, such a factory would require no outside power source and would essentially be turning depleted uranium into plutonium and antiprotons.

Suppose we used an antiproton factory specifically designed to make large quantities of antiprotons, rather than an expediently modified particle accelerator originally intended for scientific experiments (Figure 5–10). Were

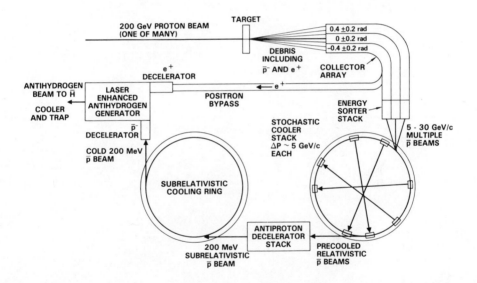

Figure 5-10 One part of an antiproton factory, a particle accelerator and associated machinery specifically designed to efficiently produce and store large numbers of antiprotons. Each beam of 200 GeV protons (there would be many) strikes a target, creating a spray of subatomic debris including antiprotons. An array of collectors channels the particles to an energy sorter stack, which allows through it antiprotons with energies between 5 and 30 GeV. Part of the array siphons off positrons. They are fed into a positron decelerator and used to produce antihydrogen. Meanwhile, the antiprotons are stochastically cooled at relativistic energies and then passed to a stack of decelerators which slow all the different energies to the same sub-relativistic energy of 200 MeV. The sub-relativistic ring cools the combined beam, which now goes to another decelerator and then to a generator that uses lasers to combine the antiprotons with the positrons and make antihydrogen. The antihydrogen beam is sent to another part of the factory where it is cooled, trapped, and stored for use.

that the case, the present energy efficiency (electrical energy out compared to mirror matter recombination energy in) could be raised from a part in a billion to a part in ten thousand, or 0.01 percent. At this energy efficiency, mirror matter would cost about 10 million dollars per milligram.

That stage of mirror matter technology hasn't yet been reached, but it's on the horizon. Some researchers are already cooking up alternate ways of producing antiprotons. Others have gone even further—on the drawing boards are the first sketches for factories that will take antiprotons and positrons and turn them into real mirror matter atoms.

Chapter Six

How to Make Mirror Matter

As we've seen in Chapter 5, current methods of making antiprotons are very costly and horribly inefficient. For high-energy particle physicists this really doesn't matter. All they want to do is use antiprotons for experimental purposes. Those who want to make practical use of mirror matter must move a step beyond the present.

Alternatives

For example, what about methods of producing antiprotons other than by high-energy protons striking a heavy-nuclei target? We know we can make antiprotons by colliding two beams of protons. The energy of collision from smashing two proton beams together is higher than that from slamming a proton beam into a stationary target. However, proton beams are not very *luminous*, as the physicists put it; there aren't a lot of protons per cubic millimeter in a proton beam. Unless methods for significantly improving the luminosity of the beams are found, the absolute rate of production will remain small.

There are ways to get around the low luminosity problem. George Chapline of the Lawrence Livermore National Laboratory in Livermore, California has proposed using colliding beams of heavy ions such as uranium with energies so high (2.5 TeV) that each nucleon has 10 GeV of kinetic energy. For a 100-meter (328-foot) diameter colliding beam facility with an intersection area of one square centimeter, the production rate could be as high as 10^{18} antiprotons per second, or a gram per week. There are problems with this concept, in part because of the large amounts of nuclear debris from the uranium-uranium interactions that would have to be filtered out. However, as design studies for antiproton factories start, heavy ion beam colliders should definitely be one of the candidate systems.

Another proposal comes from Hiroshi Takahashi of the Brookhaven National Laboratory (Figure 6-1). It has to do with the new *warm supercon-ductors. Superconductivity* is a phenomenon in which an electrical current passes through some material with no electrical resistance at all. Super-conductivity was discovered in 1911 by Dutch physicist Heike Kamerlingh Onnes. He had been studying how the electrical resistance of mercury var-ied with temperature within a few degrees of absolute zero (0° Kelvin) At a temperature of 4.2° K (degrees Kelvin) the electrical resistance of the mer-cury sample suddenly dropped to a value so low that Kamerlingh Onnes could not measure it. With no electrical resistance, a current of electricity can travel through a superconducting material with almost no loss of en-ergy. The phenomenon of superconductivity remained an interesting scien-tific oddity for many years. Then physicists began looking for ways to smash atoms and subatomic particles together with multi-gigawatts of energy. To do so with particle accelerators using standard magnets was impossible; the energy involved in creating superpowerful magnetic fields would have generated so much heat that the magnets would melt. However, magnets using superconducting coils of wire, cooled to near absolute zero with liq-

Figure 6-1   The Brookhaven National Laboratory, on Long Island, New York, is one of several locations around the world where mirror matter is made and used today. (*Courtesy Brookhaven National Laboratory*)

uid helium, could generate the intense magnetic fields needed with much less energy. It was the dawn of a new era.

However, superconduction has always been an expensive affair. Until recently, superconductivity was only known to take place at temperatures near absolute zero. Physicists needed large quantities of liquid helium for superconducting magnets. It costs a lot of money to make and store the large quantities of liquid helium. In late 1986 and early 1987, several research teams in Switzerland and the United States discovered a class of ceramic-like compounds that were superconducting at temperatures much higher than that of liquid helium. Some of these materials are superconducting at around 90° K, above the temperature of liquid nitrogen. And liquid nitrogen is cheaper by the pound than milk. A superconductor technology using warm superconductors would revolutionize research physics.

Takahashi proposes using the new warm superconductor materials in a new generation of superconducting magnets, which would produce powerful magnetic focusing fields near the region where the beams collide. This would create a tighter beam, increase luminosity, and make it possible to collect more antiprotons.

The science and technology of warm superconductors is still in its infancy. The rate of advancing knowledge in this new field, though, has been phenomenal. Already researchers at IBM and other companies have produced warm superconducting films, ribbons, and wires. It is not unreasonable to predict that the technology of warm superconductors will be far enough advanced by the mid-1990s to have a powerful impact on schemes to "mass-produce" antimatter.

Physics Versus Chemistry

In the last chapter we looked at a device for storing antiprotons called a Penning trap. Penning traps can contain a significant amount of antimatter, or mirror matter, in the form of a rotating cloud of individual antiprotons. The densities that can be reached in a well-built, high-voltage Penning trap can reach ten trillion (10,000,000,000,000) antiprotons, or about 16 billionths of a gram, per cubic centimeter. That's over a thousand times less dense than gaseous hydrogen at zero degrees Celsius, (or 273° K) but certainly a lot more dense than the best laboratory vacuum. A small, portable Penning trap with a trapping region just 5 centimeters (2 inches) in diameter can hold nearly a nanogram (a billionth of a gram) of antimatter, or nearly a quadrillion (10^{15}) antiprotons. One nanogram would easily supply the yearly needs of antimatter for medical diagnostics and cancer treatment at a major hospital, or suffice for materials inspection and atomic analysis at a large industrial facility, some uses of mirror matter we'll look at later in this book. The point here is that we don't have to wait for the mass production of antimatter atoms from antiproton ions; we can start actually using antimatter in a practical way within a few years.

As the applications for mirror matter grow, and especially as we start seriously considering the use of antimatter for propulsion, we will need to store not nanograms, but micrograms (millionths of a gram) of mirror matter. Soon the need will arise for storage of milligrams (thousandths of a gram) of mirror matter. Very large volume Penning traps have been built (some are large enough to walk into), but they are certainly not portable. We must develop a denser form of storage for antimatter. One obvious method is to store the antiprotons as some form of *antihydrogen*.

The simplest mirror atom is antihydrogen (Figure 6–2). It consists of a single positron orbiting an antiproton. The first step in making macroscopic amounts of mirror matter is to convert the antiprotons into atomic antihydrogen by adding a positron. The antiproton and positron attract each other, so the reaction takes place spontaneously. The positron jumps from one orbit (or to be more precise, "energy state") to the next on its way to the lowest orbit (or ground state) around the antiproton. As it does so it releases energy in the form of particles of ultraviolet light or *photons*. The release of these photons signals the creation of antihydrogen.

If the individual antiprotons and positrons are moving at about the same velocity, then many of them will combine together to make antihydrogen atoms. Physicists would say that their cross section is high under those conditions. If a large velocity difference exists between the antiprotons and positrons, however, then the cross section is small and the conversion rate slow. The positrons and antiprotons will probably need some outside help to combine into large amounts of antihydrogen. For example, suppose another particle is available nearby to carry off the energy released by the combination of the positrons and antiprotons. The new antiatom would then not wait to emit energy in the form of photons. Instead, it would immediately

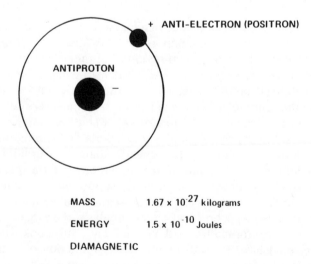

MASS	1.67×10^{-27} kilograms
ENERGY	1.5×10^{-10} Joules
DIAMAGNETIC	

Figure 6–2 An antihydrogen atom.

settle into its "ground" or resting state. This *three-body process*, would make antiatoms of antimatter much more quickly than a *two-body process*. Some other helpful techniques may include laser radiation, electrical fields, or magnetic fields. Previously formed antihydrogen atoms, ions, molecules, or clusters could also be used as catalysts or "seed particles" to enhance the conversion rate.

In the Low Energy Antiproton Ring (LEAR) at CERN in Switzerland, antiprotons are already readily available in the form of a low-energy, low-temperature beam. To make mirror matter, all that's needed is a way to combine this antiproton beam with a positron beam and have it convert into an atomic antihydrogen beam. Some scientists at CERN already carried out this experiment, but using normal matter instead of mirror matter. They did this while figuring out how to cool an antiproton beam with an electron beam. Before electron cooling was tried on antiprotons, it was first tested on normal matter protons. The scientists mixed a cold electron beam with a hot proton beam so they both traveled along in the same direction with nearly the same speed. They discovered during the tests of this "electron cooling technique" that some of the electrons recombined with the protons to produce hydrogen atoms. Theoretically, the same process would happen if a cold positron beam was mixed with a cold antiproton beam: Some of the positrons would combine with antiprotons to make antihydrogen.

European scientists at CERN have had antiprotons in abundance for years, and it is relatively easy to make positrons. It would seem likely that they would have made antihydrogen long ago. In fact, four European scientists made such a proposal right after the first low-energy antiprotons became available at the LEAR ring. The scientists proposed to produce a beam of positrons and run it side by side with the beam of antiprotons going around in their antiproton storage ring. They would then use a laser to enhance the attachment of the positively charged positrons onto the negatively charged antiprotons to produce a beam of neutral antihydrogen atoms. The neutral antihydrogen atoms would then leave the ring at the next bending magnet, while the charged antiprotons and positrons would be recirculated. The number of antihydrogen atoms made would be counted by letting them annihilate inside some detector.

When their proposal came up before CERN's experiment selection committee, though, it was given a very low priority. The reason was pretty straightforward. Every scientist knows that putting an electron next to a proton will almost automatically result in the formation of hydrogen. So doing this with mirror matter would not be research in particle physics. It would not even be physics research. At best, it would be a trivial chemistry experiment. The committee felt the antiprotons in LEAR should be reserved for more important physics experiments.

Although someone in the United States or the Soviet Union might do it earlier, the first manufacture of antihydrogen at CERN in Europe will have to wait until all the other possible "physics" experiments that can be done with a beam of antiprotons have been done first.

Other Traps

Another technique, besides that using merged antiproton and positron beams for the manufacture of antihydrogen atoms, is to start with cooled, trapped antiprotons in a Penning ion trap. A single Penning trap can store different kinds of ions at the same time, as long as they have the same sign of charge. To hold onto negatively charged heavy antiprotons and positively charged light positrons at the same time requires either two Penning traps nested one inside the other or the use of a trap that can hold and cool both positive and negative ions at the same time.

Such a trap exists. It is called a *Paul trap*.(Figure 6–3). A Paul trap uses alternating radio frequency electrical fields to form the trapping region. A Paul trap could hold both antiprotons *and* positrons and cool them until their relative velocities were low enough for the natural conversion to antihydrogen. However, efficient methods of injecting the antiprotons and positrons and removing the neutral antihydrogen haven't yet been developed. This approach still needs a lot of research before it is known whether the antihydrogen production rates are high enough to make the technique practicable.

Even if it does work, there would be additional refinements to make. After the first antihydrogen atom is formed in a Paul trap, a laser beam might be used to push an additional positron into it, to form a positively charged atomic antihydrogen ion. The antihydrogen ion could then be used in a "charge exchange reaction" with an antiproton ion. The extra positron of the positive antihydrogen ion would slip over to the negatively

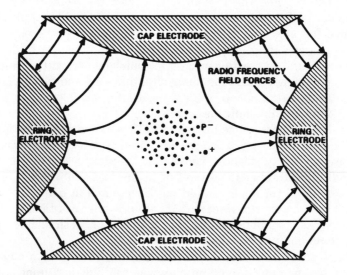

Figure 6–3   A Paul trap, like a Penning trap, can hold antiprotons without their annihilating.

charged antiproton—and form another antihydrogen atom. The positive antihydrogen ion would thus act as a catalyst to breed new antihydrogen atoms that act as catalysts for further reactions.

Holding on to Antihydrogen

Once antihydrogen has been made, it too can be slowed, cooled, trapped, and stored using a combination of electrical fields, magnetic fields, and laser beams (Figure 6-4). If a laser beam is tuned to exactly the right wavelength, the atom will absorb a laser photon. The energy in the photon will cause the electron around the atom to jump into a higher orbit. At the same time, the photon's momentum will give the atom a slight push in a direction away from the laser. The electron will soon jump down into its old orbit, emitting a photon. That photon goes off in some random direction, and in the process kicks the atom in the opposite direction. The process is repeated thousands of times per second. The reemitted photons go in all directions, and their kicks average out. The laser photons always push the atom in the same direction, so their little pushes add up. So, if a laser beam is tuned to the right wavelength, it can literally push atoms without touching them.

This laser slowing technique has already been demonstrated on beams of sodium atoms (Figure 6-5). In one experiment, researchers used a 10-milliwatt (mW) laser hooked to a special modulator that could shift the laser's light frequency up and down. The researchers began with the laser tuned well below the normal resonance frequency of the sodium atoms. This meant that the laser's light would affect only the very fast-moving sodium atoms. From the viewpoint of the sodium atoms, the laser light was shifted in frequency up to where it interacted with those atoms. These fast-moving

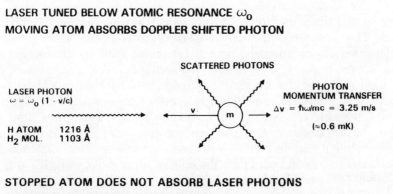

Figure 6-4   Laser cooling of neutral atoms.

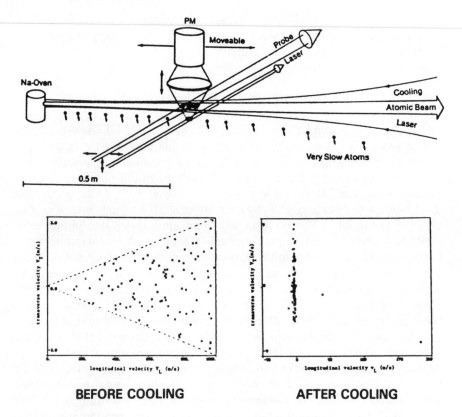

BEFORE COOLING   **AFTER COOLING**

Figure 6-5 Researchers using a tunable laser like a pushbroom were able to bring a beam of sodium atoms to a complete stop. The drawing shows the arrangement of equipment and lasers used for the experiment.

atoms slowed down and joined those atoms that were moving at moderate speeds. Then the researchers used the laser's modulator to "chirp" or tweak the laser frequency upwards. Now it interacted with sodium atoms that had a moderate speed. The fast-moving atoms that had been slowed and the moderate-speed atoms both responded to the laser light, quickly slowing down until they were all moving at low speeds. Like a pushbroom collecting marbles, the chirped laser light brought the sodium beam to a stop. The sodium beam became a slowly expanding cloud of sodium atoms with a density of about a million atoms per cubic centimeter. It was expanding at about 6 meters per second. That's the equivalent of a fast walk. It's also the equivalent of a very cold temperature.

Temperature really has to do with how fast or slowly the atoms or molecules of an object are moving: The faster the atoms or molecules move, the higher the object's temperature; the slower they move or vibrate, the lower the temperature. The water molecules in ice are moving slowly, so they cling together and form a solid. As the molecules move more and more quickly, the ice becomes a liquid. If the molecules are really jumping around, the water's temperature will be very high, the molecules will jump right out of the liquid form, and the water will become a gas we call steam. In the case of the experiment with the laser and the sodium beam, the sodium atoms were moving extremely slowly. Their velocity was equivalent to a kinetic temperature of about 50 thousandths of a degree, or 50 millidegrees, above absolute zero (0.050° K). In a follow-on experiment, the laser was deliberately over-shifted in frequency and the atoms in the beam were actually brought to a stop, then pushed back in the opposite direction.

Another technique for lining up the atomic resonance of the moving atom with the laser frequency is to use a magnetic field (Figure 6-6). This takes place inside a special solenoid as the beam of atoms barrels down the solenoid's bore. A laser with a fixed frequency slows the atoms (but does not bring them to a halt) inside the solenoid. After this initial cooling, the slowly

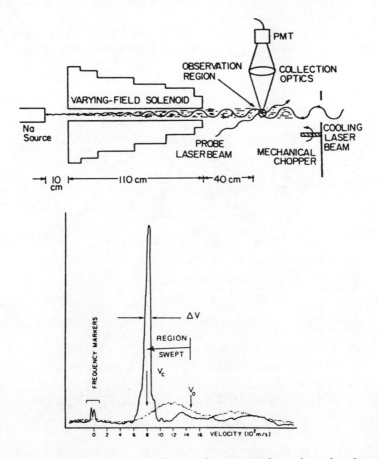

Figure 6-6 Lasers can also be used in combination with a solenoid and a magnetic field to cool and stop a beam of atoms.

moving atoms drift out of the end of the solenoid and are brought to a stop some distance away by a short decelerating pulse from the cooling laser. Scientists using this *double pulse* technique have produced a stationary cloud of free sodium atoms with a density of about 100,000 atoms per cubic centimeter and a velocity spread of about 15 meters per second, equivalent to a kinetic temperature of less than a tenth of a degree above absolute zero (0.1° K).

Trapping and Storing Antihydrogen

Lasers can be used not only for slowing, cooling, and stopping of particles, but for trapping them. One example of such an "optical trap" is shown in Figure 6-7. Two opposing laser beams are focused at the two focal points, which are located symmetrically about the trapping point. The beams spread out in radius as they travel the one-centimeter distance from the focal points to the trapping point.

The trapping point is a point of stable equilibrium. If an atom drifts from the trapping point, it meets a restoring force: the laser beam coming

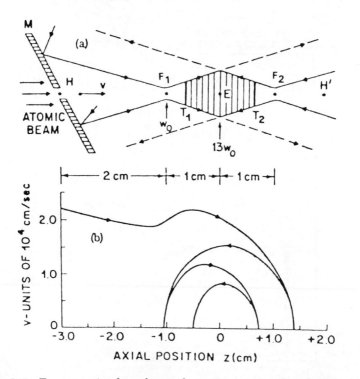

Figure 6-7 Two opposing laser beams keep atoms trapped in their focus by the pressure of the light itself. Laser traps have already been successfully used. They could be useful for trapping and holding antihydrogen mirror matter.

from opposite directions. The beam will push back any atom drifting in its direction from the trapping point. To trap the atoms after they have entered the trapping region requires *damping*. Damping occurs when the laser is tuned to a frequency below its resonance frequency with the atoms. That way the moving atoms interact more strongly with the opposing beam. The atoms enter the optical trap through a hole in the mirror. An atom entering the trap with a velocity of 200 meters per second (nearly 450 miles per hour) stops at a point about 4 millimeters beyond the second focal point. It then rebounds back and forth, oscillating about the trapping point and cooling down as it is trapped.

By the use of these traps, up to 30 million atoms can be cooled down to less than a thousandth of a degree Kelvin and be confined to a dimension as small as 1/35 of a wavelength of the laser light! At these temperatures the gas density is more like that of a solid and the atoms may convert into the condensed phase. For sodium atoms, the condensed phase is a metal and the trapping mechanism no longer will operate. However, condensed antihydrogen would be a transparent, electrically insulating material. The laser trapping will still be operative, although the details of the trapping mechanism will be different.

The atoms scientists are presently trapping are atoms of metals like sodium that absorb visible laser light. Researchers will soon be demonstrating the same cooling and trapping techniques on hydrogen. First they'll work with hydrogen atoms, and then with hydrogen molecules, which consist of two hydrogen atoms combined. This has meant developing sources of tunable narrow-band *ultraviolet* laser light, since hydrogen atoms and molecules only respond well to laser light in the short ultraviolet wavelengths. The goal will be to demonstrate the *transverse cooling* of a beam of hydrogen atoms for neutral particle beam weapons (*transverse cooling* simply refers to keeping the hydrogen atoms from drifting out of the beam at an angle). This is military-related research, of course, for the Strategic Defense Initiative (SDI), popularly known as "Star Wars." However, it will have a non-military spinoff, for once the researchers have developed these techniques, they can be used to cool and trap *antihydrogen* atoms and molecules. That will eventually lead to ways of storing and using large amounts of mirror matter for space propulsion, metallurgical testing, and even new cancer treatments.

Turning Antihydrogen Gas into Ice

Trapping the antihydrogen gas with lasers will leave the molecules at very low kinetic energies—the same as very low temperatures, a few thousandths of a degree above absolute zero. The next step is to turn the antihydrogen gas into ice by *inverse sublimation*. Sublimation is the process by which ice turns directly into a gas without going through the liquid phase—as seen when the cold gases pour off a block of dry ice (dry ice is frozen carbon dioxide; the gas is gaseous carbon dioxide sublimating off the solid ice). Inverse sublimation is simply the reverse process: turning gas directly into a solid. Normally this crystallization process occurs naturally and rapidly, especially since the laser cooling technique leaves the antihydrogen gas at temperatures well below the freezing point of antihydrogen (which is 14° K, above absolute zero). Normally this condensation and freezing process starts either on the walls of the chamber or some convenient dust particles. It's called *nucleation*, since the gas turns into a solid by condensing onto something that can act as a tiny kernel or nucleus. This is perfectly fine if the gas being turned into a solid is hydrogen gas. Unfortunately, there are no antimatter walls or antimatter dust particles to start the nucleation process for antihydrogen gas. And if the antihydrogen gas tries to condense on chamber walls of normal matter—!

Scientists need to develop techniques to induce this process of nucleation directly from the gas without involving a wall. Researchers will also have to investigate the effect of various electromagnetic fields on the nucleation process. One phase where lasers or magnetic fields could help is the precursor phase, when several molecular collisions produce "snowflake" clusters of antihydrogen molecules. The formation of the crystalline solid would follow. It's possible that laser trapping fields could set up shock waves of density fluctuations in the antihydrogen gas that would be enough to induce the nucleation. Molecular hydrogen ions might also aid in the nucleation and growth of these clusters into crystals, as might antihydrogen molecules tickled into specially selected excited states. Theoretically these techniques would work. But there is no way of knowing for sure until someone actually tries them out.

By now it is apparent that one of the critical problems with using stored antihydrogen as an energy source is the conversion of antiproton ions into antihydrogen ice. The straightforward approach is to begin by using laser photons to assist in attachment of positrons to form antihydrogen atoms. Then the atoms would be cooled down in a light trap to millikelvin temperatures and combined to form antihydrogen molecules. Then researchers would (somehow) cool and trap the antihydrogen molecules with other ultraviolet lasers. Finally, the molecules would (somehow) be formed into macroscopic balls or crystals of antihydrogen ice. This straightforward process still has a lot of "somehows" in it. Many physicists doubt that any of this could even be done, much less done economically at reasonable rates of mirror matter production.

An alternate pathway to antihydrogen ice exists, though, identified by researchers at the University of Iowa and the Air Force Astronautics Laboratory at Edwards Air Force Base in California. This approach involves the growth of condensed antihydrogen ice crystals *directly from a starting antiproton ion*. This containerless pathway to antihydrogen ice crystals uses a mechanism called *cluster ion formation* (Figure 6–8). A cluster ion is a large collection of neutral atoms or molecules that cluster around a single ion. The antihydrogen ion clusters can be quite large, with more than

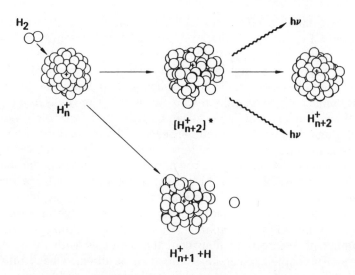

Figure 6-8 Creating and storing hydrogen cluster ions will be very important for storing and using antimatter. Growing cluster ions causes heat to be released, but after the cluster passes a threshold size, its growth is almost certain. The heat energy released as the cluster ion "seed" grows will never be enough to break up the cluster.

a thousand antihydrogen molecules per ion. Since they're ionized and thus electrically charged, they can be kept in the same Penning trap containing the starting antiproton ions, eliminating the need for building photon traps using laser beams.

Some researchers have already started looking at this cluster ion approach using normal hydrogen. Adding molecules to a cluster ion releases energy, and that *molecular association* energy is often in the form of heat. A threshold size exists for cluster ions; different paths to the creation of cluster ions have different threshold sizes. After the cluster ion passes that threshold, though, its growth is very likely. Generally speaking, the larger the "seed" cluster ion, the fewer the problems with nucleation. For example, if a hydrogen ion cluster has 30 or more molecules in it, then more hydrogen molecules can be added with impunity. The molecular association energy will never be great enough to disrupt or evaporate atoms or molecules from the cluster ion. Larger cluster ions will probably get rid of a lot of their association energy by radiating it in the infrared region of the spectrum. That will slow down the vibrations in the molecules caused by energy gained during the cluster's growth. For an ion cluster with a hundred or more molecules, individual hydrogen atoms might be added with considerable success. Until these large "seed" ions have been produced, however, nucleation rates using this pathway will be prohibitively small. The bottleneck in the containerless nucleation process is in the stabilization of the small cluster ions with fewer then 30 molecules. Our understanding of cluster formation is still pretty iffy, and needs more research and experimentation.

The problem presently under study by several scientists is to find the best way to go from a single antiproton ion to a stable cluster ion. It may take many attempts using exotic techniques and high-power special-purpose lasers to make the first antihydrogen cluster ion. The key is making the first one. When one has been made, it can be grown to large sizes, given multiple charges, and then used as a "mother seed" to "bud" smaller cluster ion seeds. The second generation cluster ions would then be used in simpler bulk growth systems. Of course, growing microgram crystals smaller than a grain of salt could only be called "bulk growth" by mirror matter technologists.

Since the process of forming the solid releases energy, it's necessary to find ways to remove the heat from the solid. A straightforward approach is to allow the solid to evaporate a molecule containing the energy from a number of fusions, and then cool the molecule with the laser cooling process. The cooled molecule would then be directed back into the cold gas. Other ways to cool the solid antihydrogen could conceivably use magnetic fields. What's important is to come up with methods that leave the antihydrogen a solid at temperatures of 1° K or lower. At these temperatures, only a few antiatoms a day evaporate from the surface.

Many paths can lead from the generation of an antiproton beam to the generation of a frozen ball of molecular antihydrogen (Figure 6-9). Example 1: The antiprotons could be trapped in a Paul trap, turned into antihydrogen atoms by the addition of positrons, trapped with photon beams, converted

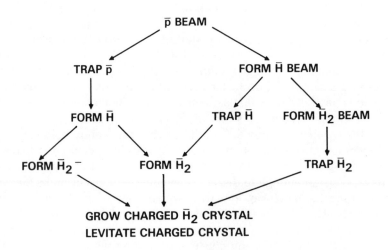

Figure 6-9 There are many paths that lead from antiprotons to antihydrogen ice. All are theoretically possible. Any one of them, once mastered, will lead to the creation and storage of true mirror matter in the form of antihydrogen ice crystals.

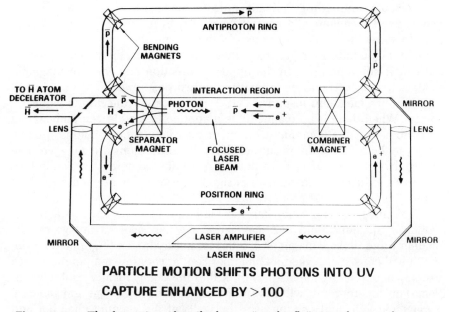

PARTICLE MOTION SHIFTS PHOTONS INTO UV
CAPTURE ENHANCED BY >100

Figure 6-10   The formation of antihydrogen "on the fly" using beams of antiprotons (top) and positrons (bottom). The beams are slowed using lasers, combined in a magnet (right), and then converted, using a laser, into antihydrogen. Another magnet (left) separates the uncombined positrons and antiprotons from the antihydrogen. The antiparticles circle back into their rings and come around again for another combination attempt. The antihydrogen beam continues on to a decelerator ring and is eventually cooled and stored as antihydrogen ice.

into magnetic field-trapped antihydrogen molecules, tickled into forming an iceball by positron beams and laser beams, then grown into a charged microcrystal. And Example 2: The antiproton beam could be converted into an atomic antihydrogen beam, then a molecular antihydrogen beam, all on the fly (Figure 6–10); then the antihydrogen molecules could be slowed, cooled, trapped, and turned into ice with ultraviolet laser beams. A few extra positrons would prepare the crystal for levitation in an electrostatic trap.

One way or another, scientists will eventually succeed in making solid antihydrogen. So let's assume that one of these techniques has succeeded. Macroscopic amorphous or crystalline balls of antihydrogen ice have been made. They should be at a very low temperature to prevent evaporation and large enough to scatter infrared light so their position can be determined. They will mass from nanograms to milligrams.

Levitating Antihydrogen Ice

Once the small microcrystals or larger ice balls of antihydrogen ice are formed, they can be transferred to a compact lightweight storage and transport system that uses simple magnetic or electric traps for levitation.

Molecular antihydrogen, like its mirror cousin, molecular normal hydrogen, is nearly magnetically neutral when it is in its lowest energy state or ground *state*. The two antiprotons and the two positrons each have a magnetic moment, but in the ground state of the molecule, the two antiprotons have their spins pointing in the opposite direction and the two magnetic fields cancel out. The same is true for the two positrons.

The only magnetic response that is left is called the *diamagnetic* response. When there is no magnetic field applied to a hydrogen molecule, it has zero net *magnetic moment*. When an external magnetic field is applied, however, the orbits of the positrons about the antiprotons are changed. The changed orbital motion is equivalent to an additional current, which in turn causes an induced magnetic field of sign opposite to the applied field. This induced magnetic field property is called diamagnetism.

A diamagnetic molecule has the tendency to move toward a region with low magnetic field and stay there. We can take advantage of this tendency and make a magnetic "bottle" for antihydrogen (Figure 6–11). The bottle will consist of a vacuum chamber kept at very low temperature, with two superconducting metal rings built into the cryogenically cooled walls. Persistent supercurrents will flow in the superconducting rings and generate a magnetic field that's strong near the walls of the vacuum chamber and weak at the center of the chamber. Any antihydrogen put into this container will avoid the strong magnetic fields near the walls and conveniently *collect in the center* as a ball of antihydrogen ice. This passive magnetic levitation technique is the preferred suspension method for antihydrogen. It would be safer than any active levitation technique, as well as making the storage and

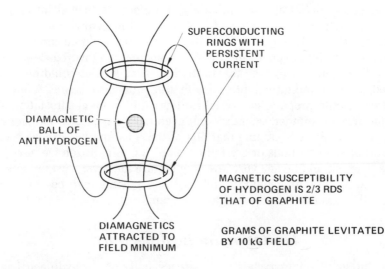

SUPERCONDUCTING
RINGS WITH
PERSISTENT
CURRENT

DIAMAGNETIC
BALL OF
ANTIHYDROGEN

MAGNETIC SUSCEPTIBILITY
OF HYDROGEN IS 2/3 RDS
THAT OF GRAPHITE

DIAMAGNETICS
ATTRACTED TO
FIELD MINIMUM

GRAMS OF GRAPHITE LEVITATED
BY 10 kG FIELD

Figure 6-11   A magnetic "bottle" will hold a cold ball of antihydrogen. The magnetic fields keep the ball levitated in the center of the "bottle."

transport container extremely simple, compact, and independent of electric power.

Not only can we use magnetic fields to store and manipulate antimatter, we can also use electrical fields. If a ball of antihydrogen ice has a slight excess electrical charge, electrical fields can move it around. For example, a weak beam of ultraviolet light will drive a few positrons off the antihydrogen ice ball and keep it electrically charged. Researchers at the Jet Propulsion Laboratory have been experimenting with such traps for use in zero gravity laboratories on Spacelab and the Space Station. In fact, they have already used such traps to levitate 20-milligram balls of water ice (Figure 6-12).The density of water (one gram per cubic centimeter) is 13 times that of antihydrogen ice. So the JPL trap—which exists today—could levitate a 1.5 milligram ball of antihydrogen ice of the same size, surface area, and surface charge of the water ice ball. What's more, it could be done in a space vehicle traveling at an acceleration of 13 times the force of Earth gravity.

The JPL traps consist of two large curved metal plates about 10 centimeters (4 inches) apart in a vacuum chamber. Between the two plates is a ball of ice that is kept charged by ultraviolet light that kicks electrons off the ice and leaves the ice ball positively charged. The top plate of the trap is charged negatively and the bottom plate is charged positively, so the ice ball is levitated in the Earth's field by the electrical fields from the charged plates. The curvature of the two plates keeps the ice ball centered radially.

The vertical position of the ice ball is not stable. If the ice ball approaches the upper plate, the attraction of the charges on the plate becomes stronger. Thus, unless prevented, the ice ball will quickly slam into the upper plate. In the JPL trap, the voltage between the plates is varied by a fast-acting voltage generator controlled by a television camera "watch-

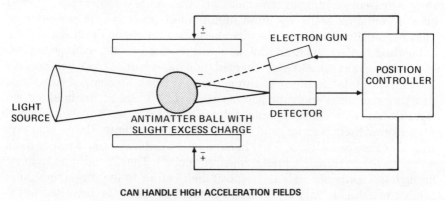

CAN HANDLE HIGH ACCELERATION FIELDS

JPL LEVITATED 20 mg ICE BALL IN 1 gee

Figure 6-12 A charged ball of antihydrogen ice can be kept suspended in a vacuum between two electrically charged plates.

ing" the position of the charged ice ball. The camera controls the voltage generator so the ball stays levitated in the center of the chamber. This is an example of an *active* levitation technique. It works quite well. Its only drawback is that if the mechanism fails the iceball slams into one plate or the other. That's not a problem with normal ice, but it is something one probably doesn't want to happen with a ball of antihydrogen ice. For safety reasons, some people will prefer *passive* levitation techniques.

Storing Antihydrogen Ice

By now it's apparent that it should be possible someday to form and levitate antihydrogen ice. Laser cooling will leave the antihydrogen at a temperature well below a millidegree, and the antihydrogen ice will start out cold. The vapor pressure of antihydrogen at $1°$ K (one degree above absolute zero) is so many orders of magnitude below any pressure that has ever been measured that it is effectively zero. So if the antihydrogen can be kept below this temperature there is essentially zero sublimation loss. The vapor pressure rises very quickly with the temperature, however. At $2°$ K it has risen to a point where the sublimation rate is about 1000 antiatoms per second from a milligram-sized ball. At $4°$ K sublimation of antihydrogen atoms from the solid is even more rapid. To keep the sublimation rate of the antihydrogen down to controllable levels, we need to keep its temperature below $2°$ K. The walls of the storage and transport container also have to be cooled well below liquid helium temperatures to keep the walls from *outgassing* (that is, to keep gas molecules trapped in the walls from "trickling out" into the container vacuum).

Even if the antihydrogen ice and the container walls are kept cold, there will always be a few stray antihydrogen molecules leaving the ice ball and annihilating on the container walls. There will also always be a few stray outgassing hydrogen molecules or other normal molecules knocked off the container walls by those annihilation processes or cosmic rays. They're going to drift across the chamber to annihilate on the surface of the antihydrogen. Each such annihilation will produce on the average six 200 MeV gamma rays, six 250 MeV charged pions, and four 511 keV gamma rays from the annihilation of the positrons and electrons. So charged subatomic particles and some energetic gamma rays will always be flying about, no matter what we do to keep things (literally) cool.

However, there is some good news in this state of affairs. The gamma rays will not cause any intense local heating in the antihydrogen. Most of them will pass right through a milligram-sized iceball. They'll continue right on through the container wall and deposit their energy in the container's outer gamma ray shield. Only a very small amount of energy is deposited in the antihydrogen iceball. Even when an annihilation takes place right on the surface of an antihydrogen iceball, the event will only deposit 5 trillionths of a joule of energy—that's the equivalent of a trillionth of a watt of heat input for two annihilations per second. No one will ever notice, including the antihydrogen iceball.

Portable Refrigerators

To keep solid antihydrogen a solid for long periods of time, it will have to be stored at temperatures well below that of liquid helium. To prevent heating by radiation from the surroundings, the walls of any trap must be kept cold, preferably well below 1° K. Such temperatures used to be very difficult to attain. They required specially constructed refrigerators that used helium-3 and helium-4, or that used a special form of salt called *paramagnetic salt.*

However, things have changed in recent years. One driver of change has the initials "SDI." Researchers working to develop sensors for the Strategic Defense system have been very interested in compact cooling. They want to keep orbiting infrared detectors and optics as cool as possible. That has led to the development of compact, continuously operating magnetic refrigerators, which will probably be available when needed for future research in the storage of antihydrogen.

These new kinds of refrigerators are relatively simple. A slowly rotating wheel carries blocks of paramagnetic crystals such as gadolinium-gallium-garnet. These blocks of magnetic material first move into a high magnetic field region usually supplied by a persistent current superconducting coil. There the material is magnetized. The process of magnetization also heats the material. The block of magnetic material then releases its heat into a "high" temperature bath of liquid helium, which is at 4.2° K. That cools it to the temperature of the liquid helium. The magnetic material continues rotating out of the magnetic field region to the cold loop heat exchanger where it demagnetizes. That in turn cools the inner wall of the levitation chamber, which holds the levitated ball of antihydrogen ice, to just 1° K.

Present models of these super-refrigerators typically weigh less than 70 kilograms (150 pounds), and they can already cool material to temperatures as low as 1.5° K, a mere one-half degree above the target temperature. Different magnetic materials can allow even lower temperatures to be reached. Instead of using liquid helium evaporation to maintain a 4.2° K heat bath, scientists could build a closed-cycle refrigerator that would eject heat at room temperature. This completely closed, continuously operating system would require about 2,000 watts of input electrical power to obtain one-half watt of cooling at 1° K—the target temperature.

Removing Antimatter from Storage

Getting solid antihydrogen into a container and keeping it there is a feat in itself. But if we want to use the antihydrogen, we need also to be able to remove it from the container without annihilating it, and then direct it into a rocket engine. Several different techniques exist to do this. Let's suppose the antihydrogen is in the form of an electrostatically suspended ball many milligrams in size. First an ultraviolet laser is used to drive some positrons off the iceball. Then an intense electrical field is used to remove some of the excess antiprotons and send them to the rocket's thrust chamber.

However, it might be more desirable to form the antihydrogen as a cloud of charged microcrystals. Each crystal would be a microgram in mass and contain the energy equivalent of 20 kilograms (44 pounds) of chemical fuel. In that case, a directed beam of ultraviolet light would be used to drive off a few more positrons from one microcrystal. Then that individual microcrystal could be pulled from the microcrystal cloud using electrical fields, and directed down a vacuum line to the thrust chamber. Since the position of the charged microcrystal in the injection line can be sensed, mechanical vacuum shutters can allow the passage of the microcrystal without breaking the storage chamber vacuum.

It will be many years before mirror matter becomes a product that is bought and sold like gasoline or diesel fuel. There are myriad difficult problems left to solve. The production, capture, cooling, storage, transport, and use of this extremely potent, extremely expensive, nearly magical source of raw energy will take time, money, and a lot of hard work. Skeptics doubt it will ever happen. But others have been looking closely at the problems that have already been overcome and the problems still left to be solved. It is now starting to appear that there are no impassible barriers on the road to the creation of a mirror matter technology. Space propulsion with mirror matter is now a distinct possibility. No physical laws will be violated. Mirror matter, in some form, will someday be made and stored in enough quantity to produce the megawatts and gigawatts of prime power and propulsion needed for rapid space travel.

And when we say rapid we mean *fast*. On January 24, 1986, the Voyager 2 space probe made the first-ever flyby of the distant planet Uranus. It had taken Voyager 2 nearly eight years to get from Earth to Uranus. A few days later, a high NASA official said it would be another 200 years before any robot space probe returned to Uranus.

Your not-so-humble authors disagree. We're convinced that humanity will return to Uranus much sooner than two centuries from now. We think humans—not robots, *humans*—will look upon Uranus with their own eyes. We think this will happen within the lifetime of some of you readers.

One of those humans may well be you. And you will get there with antimatter propulsion.

Turn Left at the Moon: IV

Dyson City had the shape of a huge rotating wheel. There was a basic framework of open girders about 200 meters in diameter rotating about once per minute to produce enough gravity to give a sense of up and down. Two opposite quarters of the wheel had been filled in with a curved pressure hull 15 meters in diameter. To Eric, Dyson City looked like a doughnut with two large bites taken out of it. The spaceplane docked at the center hub of the wheel. As they prepared to disembark a sudden flare of light in space caught Eric's attention.

"Hey, Dad, what was that?" he asked.

"Don't know," Jim replied.

"Hello," said the captain's voice. "For those of you wondering about the flash of light, I've just been told that was a Brazilian military spaceplane based at Helin. They're heading out on an emergency mission to the Japanese polar station. They've had a serious accident out there, and have asked for medical assistance. The Brazilian ship will rendezvous with the space station and check out the casualties. If it's really bad they'll shuttle them downside. Otherwise, they'll bring them back here for treatment. Just thought you'd like to know."

Jim shrugged. "Now we know. Good thing they have antimatter propulsion systems for those babies. They wouldn't be able to pull that mission off without them."

The passengers finished disembarking. Eric, his parents, and his grandmother took an elevator down to the A sector. As the elevator descended they all floated to its floor and began standing on their feet again.

They were now at the upper deck of the A sector. Four meters above them was a huge curved window that looked inward to the hub and beyond to the curved window of the C sector. Eric looked up. People were hanging upside down above him, in the C sector, looking through their ceiling at him.

A large mirror to one side of the o'neill reflected the sunlight into the windows and down through light-shafts to the farms on the decks below. They had a day before they would board the Mars Longliner, so there was time to explore Dyson City. Jim headed for C sector to visit the Russian cosmic ray scientists there. Ginny took off for the lower decks of A sector to visit the hydroponic garden.

"Well, Red, it's you and me. Want to see some construction?"

Eric shrugged. "Sure." The prospect of tagging along after his mother was not terribly exciting.

They headed for B sector. Although all the sectors of the city were open to everyone, the A sector had been primarily built by Americans, and held their experimental facilities. The Russians had built C sector. The European Space Agency was expanding B sector, and since Kirsten worked for ESA, she had access to some areas that were not on the tourist maps. She and Eric stopped in at a control center to look out a window to watch the work. Construction workers in hardsuits with powerful propulsion units and manips were assembling various components into a new section of the sector.

As they entered the control room, they saw the workers bringing up some titanium alloy plates. The plates would get one last inspection before being welded into place to make the bottom of the pressure hull—the "bedrock" of the huge curved structure. The man at the console looked up as they came in. "Hey, Dr. Docherty, good to see you again."

"Hi, Andre, how's Trix?"

"She's fine. This your boy?"

Kirsten introduced Eric to the man, who it turned out had worked with her for a year at Turin before leaving to work for Bechtel Interplanetary.

Andre excused himself and turned back to his console. A two-by-three meter plate was being floated in front of a p-bar scanner. The workers swept a beam of antiprotons across the five-centimeter-thick plate. Inside the control room, Eric and his mother watched the image build up on a screen in front of Andre. In ten seconds the scan was finished. To Eric it looked like a dull gray rectangle. Other eyes were keener.

"Look," said Andre as Kirsten said, "There, in the lower left corner." At that moment the computer dropped a red rectangle on the same area. Andre rotated and expanded the image, then stretched the contrast. "We have a flaw here, folks," he said, and reached for the "Reject" key. "We don't want a flawed plate in the bottom of Dyson City. If that failed," he was explaining to Eric, "the results would be catastrophic."

"No, wait," said Kirsten. "It doesn't seem that bad. I don't see a void there."

"Hmmm. True. True. It's a collection of line dislocations. Well . . . yes. We can anneal them out." Andre pushed the "Rework" button. Outside, a worker waved a manip in acknowledgment. A small robot sprayed a prominent white mark on the plate, and it was towed off to a holding area. There a separate team of workers moved their heavy-duty antiproton beamer into position and turned it on. Within moments, a microgram of antimatter heated up the local flawed region inside the plate and annealed away the dislocation flaws. The annealed plate would not be used for the pressure hull; safety regulations forbade that. But it would still find a use in Dyson City. A worker retagged it with yellow paint and moved it into a storage orbit.

Meanwhile, another plate was moved into position and scanned. Andre's console signaled it as fine. The construction workers moved it into position, ready for welding. Dyson City continued to expand.

Eric decided that tagging after his mother *was* rather interesting after all.

Chapter Seven

Rocket Power and Mirror Matter

Who among us has not been thrilled by the sight of a rocket launch? Perhaps some of us have actually witnessed such an event firsthand: a Space Shuttle launch, or one of the Saturn/Apollo moon rocket launches back in the 1960s, or even just the launch of a small rocket built by a local amateur rocket club. Or maybe we have only seen a rocket launch on television. It doesn't matter; the event is an exciting one, whether seen directly or through the electronic medium. Rocket launches have become so common today they're almost invisible. Until the tragic loss of the Challenger in 1986, the TV news media had pretty much ended live coverage of space launches. Editorial cartoonist Jeff McNelly summed it up in a cartoon he did several years ago. It shows two old codgers sitting in beach chairs under the California sun. Overhead a giant Space Shuttle swoops down towards a landing. The first codger is saying to the second: "Well, there goes the ol' Four Twenty-Three."

As of this writing, "the Four Twenty-Three" is canceled; U.S. Space Shuttle launches will not resume until mid-or late 1988. However, the Soviet Union's space program has continued to move forward at a steady pace. The U.S.S.R. routinely launches more than a hundred manned and unmanned satellites yearly.

This routine matter of firing rockets into orbit around the Earth or sending space probes to other planets is a very new capability for humanity. The Space Age only began on October 4, 1957, when the Soviet Union put the first artificial satellite (Sputnik) into orbit. But we've learned quickly. Just 12 years later the human race had gone from placing its first small object in orbit around Earth, to placing two of its representatives on another heavenly body.

Rockets work because Isaac Newton's Third Law of Motion is more than just a bunch of words: "For every action there is an equal and opposite reaction" (Figure 7–1). Put on a pair of roller skates. Take off your watch. Throw your watch out in front of you. You will glide backwards. Action—reaction. Strap yourself into the commander's seat of the Space Shuttle. Fire the rockets. Millions of kilograms of solid and liquid fuel burn and turn into fiery exhaust streaming out the rocket nozzles. You and the shuttle leap off the pad and streak up and out towards space. Action—reaction. Now you are in orbit. You fire the shuttle's maneuvering rockets. The exhaust gets thrown out the nozzles in one direction. The shuttle moves in the other direction. Action—reaction. You do not move in that direction because the rocket exhaust is "pushing" against something—in the vacuum of space there isn't anything to push against. You move because of Newton's Third Law of Motion: For every action, there is a reaction. In essence, rockets are pretty simple. In actuality, of course, they are not. Other factors come into play.

Rockets have always been *theoretically* possible; the first ones were built by the Chinese hundreds of years ago. However, it has only been in the middle and latter parts of the twentieth century that space-traveling rockets have been *technically* possible. Rocket pioneers like Konstantin Tsiolkovsky, Hermann Oberth, and Robert Goddard first developed the mathematical tools and early engineering techniques that eventually made space travel via rocket a commonplace event. In doing so they introduced

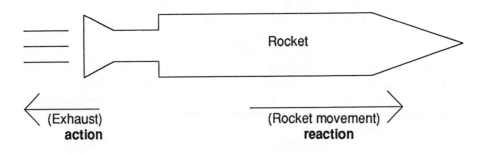

"For every action there is an equal and opposite reaction."
—Isaac Newton

(Exhaust)
action

Rocket

(Rocket movement)
reaction

Figure 7–1 Action and reaction is the basic law of rocketry. The action of exhaust gases streaming out of a rocket in one direction sends the rocket in the other direction (reaction).

the world to some heretofore unfamiliar concepts: *thrust, exhaust velocity, specific impulse, delta-v, mass ratio* and *staging*. Let's take a look at them.

Thrust and Exhaust Velocity

Thrust is a measure of a rocket's propulsive force. It depends solely on the quantity of rocket exhaust gases produced and the velocity with which these exhaust gases stream out of the rear of the rocket. Thrust is measured in pounds (actually pounds-force) by engineers, while scientists measure thrust in *newtons*. One newton is equal to 0.22 pounds, and conversely one pound is equal to 4.5 newtons. Each of the three main engines of the U.S. Space Shuttle can develop up to 106,000 pounds of thrust, for a total of nearly 320,000 pounds of thrust. The Shuttle's solid rocket boosters (or SRBs) each develop about 550,000 pounds of thrust. Thrust gets talked about a lot, because the big numbers sound so impressive. Certainly a booster with 550,000 pounds of thrust sounds more powerful than an engine with a tenth of a pound of thrust. However, the engine with a tenth of a pound of thrust may be an ion engine. As we'll see in Chapter 10, that kind of space propulsion system may be able to send a space probe to Pluto. The Shuttle's SRB most certainly cannot. The reason is *exhaust velocity*.

The faster a pound of reaction mass comes out of the rocket nozzle, the greater the propulsive thrust that pound will give to the rest of the rocket. A fuel with a high energy content per pound has a greater exhaust velocity when it comes out of the nozzle and provides a higher thrust. Gunpowder rockets have a low energy content per pound of powder and have an exhaust velocity of about one kilometer per second (km/s). Solid storable fuels such as those used in the solid rocket boosters and in orbit transfer rockets to take payloads from low earth orbit (LEO) out to geosynchronous earth orbit (GEO) use higher energy density chemicals and achieve exhaust velocities of about 3 km/s. The most powerful fuel presently used in the space program is a mixture of liquid hydrogen and liquid oxygen that is burned to make water vapor, and used in the Space Shuttle main engines. It has a theoretical maximum exhaust velocity of 5 km/s. The designers of the Shuttle's main engines have done a remarkable job optimizing their performance. They get an exhaust velocity of 4.5 km/s, or 90 percent of the theoretical maximum performance of the fuel. There are a few dangerous chemical combinations with slightly higher energy densities than liquid hydrogen and liquid oxygen. One is liquid fluorine and liquid hydrogen that burn to form hydrofluoric acid vapor. This fuel has a theoretical maximum exhaust velocity of 7 km/s, but not surprisingly the engineers at Kennedy Space Center don't want to use it! There are no chemical combinations that give a higher exhaust velocity (all the chemical combinations are known, and the exhaust velocities obtainable from these fuels are known; see Table 7–1). Chemical fuels have reached their limit and new fuels must be found. One such new fuel is, of course, antimatter.

Table 7-1   Chemical Space Propulsion Fuels

Fuel	Oxidizer	Approximate Exhaust Velocity (km/s)	
Kerosine	Red fuming nitric acid	2.5	*
Kerosine	Liquid oxygen	3	**
Hydrazine	Chlorine pentafluorine	3	*
Hydrazine	Liquid oxygen	3	**
Pentaborane	Hydrazine	3.2	*
Liquid hydrogen	Liquid oxygen	5	**
Liquid hydrogen	Liquid fluorine	5.2	**

*Storable propellants; can be stored in the rocket at "room temperature."

**Cryogenic propellants; either oxidizer or fuel, or both, must be stored at extremely low temperatures.

Specific Impulse

Some engineers measure the performance of a fuel in terms of its *specific impulse*. The term "specific" in engineering parlance means "per pound," so specific impulse is the amount of impulse per pound. In a rocket engine, that means the amount of impulse or force that is obtained per pound of fuel. Unfortunately, the term "pound" in engineering units means two different things. First, pound is used to describe the mass of something, the quantity of a substance. Second, pound is used to describe force. The two are easily confused, since on the Earth, where the gravity field is always one G, a pound of mass produces a weight of a pound, and weight is a force.

Thus, when early rocket engineers were trying out different fuels they found that some fuels gave greater thrust per pound of fuel used per second than other fuels. To quantify this performance they defined the quantity called specific impulse, or the amount of thrust in pounds (force) from a specific unit mass of propellant used in pounds (mass) during a specific unit of time. The proper units of specific impulse are thrust force divided by mass flow per second, or pounds(f)/pounds(m)/sec. Since to the early rocket engineers, a pound was a pound, they canceled one type of pound out with the other, and expressed the specific impulse in seconds.

For example, rocket engine A can produce 6 million pounds (force) of thrust by burning 20,000 pounds (mass) of propellant per second. It therefore has a specific impulse of 300 seconds. Rocket engine B, on the other hand, produces 18 million pounds of thrust with the same amount of propellant used per second, and has a specific impulse of 900 seconds. Obviously, rocket B is using a propellant which is much more powerful than the propellant used by rocket A, and is therefore a more efficient rocket engine.

To give a real-life example, the Space Shuttle's main engines have a specific impulse of 460 seconds in a vacuum. That's about the best specific impulse ever achieved with a conventional chemical-fuel rocket. A specific impulse of 900 seconds is unheard of for such a propulsion system. No known chemical fuel has that much energy per pound. Nuclear rockets or arcjet propulsion engines, however, are able to achieve such specific impulses. And ion propulsion engines, which have thrusts measured in tenths of pounds, can easily achieve specific impulses much higher than even that. They use an external source of energy to eject the propellant mass. Their specific impulses end up being very high, often in the thousands of seconds. Though they have a very small total thrust, their long thrusting time more than makes up for it.

Specific impulse can easily be converted into the equivalent exhaust velocity by just multiplying the specific impulse by g, the acceleration of gravity, in the proper units. That's 9.8 meters per second per second on Earth. A specific impulse of 500 seconds therefore translates into an exhaust velocity of about 4,900 meters per second or 4.9 km/s.

Δv and Mass Ratio

Almost as famous as Einstein's equation relating mass and energy is an equation worked out by the early rocket pioneers. It relates the velocity achieved by a rocket in free flight (that is, no air and no gravity) to the velocity with which the exhaust leaves the rocket. We might call it the "rocket equation," since it tells us something very important about making a rocket work. Or we could call it the Δv equation, since that's the result of working it out. v stands for velocity. Delta (Δ) is a letter in the Greek alphabet that rocket scientists and engineers use to mean change. Delta-v is "velocity change." In mathematical terms, delta-v is equal to the exhaust velocity multiplied by the natural logarithm of the number that results from dividing the rocket's initial (fueled) mass by its final (empty) mass. It looks like this:

$$\Delta v = V_e \times L_n \ (M_i/M_f)$$

V_e is exhaust velocity; L_n is the natural logarithm; M_i is initial mass; and M_f is the rocket's final mass, after the fuel is burned up. This equation may seem scary; it is not. You can use it even if you don't know what "logarithm" means.

The ratio of initial mass to final mass is called the *mass ratio*. When the initial mass equals the final mass, their ratio is 1. (Any number divided by itself equals 1.) The rocket has not used any fuel. The natural log of 1 is 0. If the initial mass is twice the final mass, then the rocket has obviously burned up half of its initial mass (the fuel). The rocket's mass ratio is 2, and the natural log of 2 is 0.69. The "rocket formula" says that this rocket will achieve a velocity (Δv) that is 0.69 or 69 percent of its exhaust velocity (V_e, remember?). If the rocket burns more fuel, its mass ratio will get larger. The formula shows that it will therefore go even faster. Table 7–2 describes this relationship. There are also several other ways to determine a spacecraft's mass ratio (Table 7–3). We can figure it out if we only know the rocket's *mission characteristic velocity* (the total of every change in velocity that takes place during the mission) and its exhaust velocity. Or suppose we know the specific impulse of a rocket instead of its exhaust velocity. We can multiply the specific impulse by "g," (9.8 meters per second per second) to learn the exhaust velocity, and then figure out the mass ratio.

The larger the difference between vehicle mass and propellant mass, the larger the mass ratio. And the larger the mass ratio, the less payload you can move into space or around the solar system per given amount of fuel. The rocket equation is a law of diminishing returns. If you want to go faster, it is more effective to increase *exhaust velocity* than to keep adding fuel. This makes sense if you consider the fact that the rocket must carry its fuel with it. The more fuel it carries, the more fuel it needs to carry the fuel!

If the rocket is taking off from the ground, it needs to expend fuel fighting gravity and air resistance as well as building up velocity. In fact, a rocket going from Earth to near-Earth orbit has to work as hard as a rocket achieving a velocity of 10 km/s in free space, even though orbital velocity is only about 7.5 to 7.7 km/s (depending on the altitude of the orbit). In essence, the mission characteristic velocity for an Earth-to-orbit mission is about 10 km/s.

Table 7-2   Mass Ratio and Exhaust Velocity

If mass ratio is:	the multiple of exhaust velocity is:
1	0.00
2	0.69
2.718 ("e")	1.00
4	1.39
6	1.79
8	2.08
10	2.30

Table 7-3   How to Figure Mass Ratio

$$R = \frac{m_v + m_p}{m_v} = e^{V/v}$$

"R" is mass ratio. "m_v" is the mass of the vehicle. "m_p" is the mass of the propellant. "V" is the mission velocity. "v" is the rocket's propellant exhaust velocity, and "e" is 2.718.

Staging

No space vehicle currently in existence can go straight from the ground into orbit in one jump. All orbital launch vehicles use the so-called *staging system*. A rocket's payload consists of one or more smaller rockets, usually stacked one on top of the next. The payload of the rocket's first stage is the second stage rocket. The second stage in turn may have a third rocket (or stage) as its payload, and so on. Once the rocket engines in the first stage have used up their fuel and have shut off, the second stage's engines fire up. The first stage drops away and falls back to Earth. Meanwhile, the second stage is already traveling at a high speed when its engines start. It can reach a much higher velocity than if it were starting from scratch. This process continues, stage by stage, until the final stage with its payload reaches the speed it needs. Every space launch vehicle ever built to put payloads into orbit has been a staged rocket.

The Space Shuttle is a two-stage spacecraft. Its strap-on solid rocket boosters are one stage, and liquid oxygen/hydrogen engines are the second stage. The shuttle has an exhaust velocity of about 4.5 km/s when it lifts off. In order to get into orbit a single-stage shuttle would have to carry 8.2 times its weight in fuel, for a mass ratio of 9.2. That's just too hard to do. The delicate Voyager airplane that made the famous trip around the world without refueling had a mass ratio of about 5. And it was almost all fuel tank when it took off. The shuttle has a few tricks, though: the two large solid rocket boosters and the expendable external fuel tank. The solid booster stages do the work of fighting air resistance and gravity. That means the shuttle can lift off with lighter engines and only needs to get about 7.5 km/s out of its liquid oxygen/hydrogen system. It carries this liquid fuel outside the shuttle in the external tank, so the shuttle doesn't need landing gear and wing area to land the fuel tank mass. The shuttle orbiter itself, with its liquid oxygen/hydrogen propulsion system, has a mass ratio of about 5, so it can carry a decent-sized payload in the payload bay. But throwing away the external tank and recycling the solid boosters costs a lot of money—shuttle missions end up costing about $100 million per shot.

Single-stage-to-orbit launchers would need propulsion systems much more fuel-efficient than what we have today. One proposal, the *aerospace plane*, would use a combination of air-breathing engines and high-energy chemical rockets. But in a sense, even this is a "two-stage" vehicle, since it uses two different kinds of propulsion systems. For now, orbital launch vehicles will continue to use the staging system.

Antimatter Rockets

The idea of an antimatter-powered rocket may seem like science fiction. But the truth is that such rockets have been seriously studied on and off since the end of World War II. The first studies of antimatter rockets were carried out in the early 1950s by the German rocket scientist Eugene Sänger. At the time Sänger wrote about his antimatter rocket, the only known antimatter particle was the positron, which could be obtained from the decay of certain radioactive elements. (The antiproton, as we saw in Chapter 2, was not discovered until 1955.) The interaction of electrons and positrons results in the annihilation of both particles and the generation to two energetic gamma rays. If the particles had low kinetic energy before their mutual annihilation, the two gamma ray photons would each have an energy of 511,000 electron volts, or 511 keV. The wavelength of the gamma rays would be very short, some 100,000 times shorter than regular visible light photons. But gamma rays are nevertheless photons, light particles. And since the products of the annihilation of positrons and electrons were two high energy gamma ray photons, Sänger called his antimatter rockets "photon rockets."

The gamma rays from the annihilation of electrons and positrons come flying out in every direction. In order to use them for propulsion, Sänger needed to find a way to build a mirror to direct the gamma ray photons out the rear of the rocket to obtain thrust. Gamma rays have so much energy, however, that they penetrate deeply into any known material. Sänger spent the rest of his life, trying to find a method to make a dense electron-gas "mirror" to direct those gamma rays and move his "photon rocket" from science fantasy to science possibility. He never found it.

In 1955, Emilio Segrè and his colleagues discovered the antiproton at the Lawrence Berkeley Laboratory. As they and other researchers created more and more antiprotons, they learned more about the process of proton-antiproton annihilation. They discovered that the annihilation of antiprotons with protons does not simply produce gamma rays; the process is much more complicated. When researchers first learned this in the 1950s, it was scientifically interesting but not much else. Today that "scientifically interesting" information promises to be utterly revolutionary. The antiproton's complex annihilation process makes it much more suitable than the positron for antimatter propulsion systems. Sänger's "photon rocket" may never be built. Someday soon, though, someone is going to build a "pion rocket."

The annihilation of an antiproton with a proton does not produce gamma rays immediately. Instead, the annihilation produces several short-lived elementary particles called pi-mesons or pions. On the average, a proton-antiproton annihilation first kicks out 3.0 charged pions—1.5 positively charged pions and 1.5 negatively charged pions—and 2 neutral pions. The paths of the charged pions can be bent into curves by a strong magnetic field. This means that the charged pions can be partially contained, controlled, and directed by magnetic fields. Also, the charged pions can heat up a liquid (such as liquid hydrogen) as they pass through it. The electrical charge on the pions transfers some of their kinetic energy into heat energy in the liquid. If enough annihilations were taking place, the liquid would be heated until it became a high-pressure gas. And that gas in turn could be used to propel—a rocket.

How an Antimatter Rocket
Works—Two Versions

Let's look inside an antimatter rocket (Figure 7-2) where antiprotons and protons are being annihilated. The very short-lived neutral pions from the annihilation process almost immediately decay into two gamma rays. However, the charged pions have a normal half-life of 26 nanoseconds (26 billionths of a second). Ah—but they are moving at 94 percent of the speed of light. And at that speed *time slows down for them*. It's Einstein's Theory of Special Relativity again. The Einstein time dilation effect lengthens their lives to 70 nanoseconds. Of course, that's still a very short lifetime, but 94 percent of light-speed is *very fast*. In one second they would travel nearly 282,000 kilometers (169,000 miles). Even with their very short lifetime, they move pretty far. The average distance traveled by the charged pions is 21 meters (60 feet). That's a very long distance compared to the size of, say, a typical rocket engine.

By using a strong magnetic field outside the engine, the charged pions

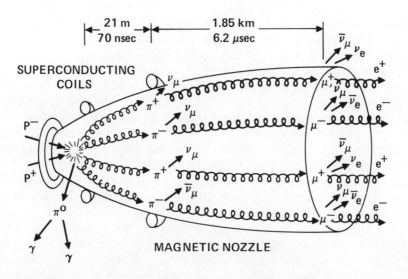

Figure 7-2   This diagram of a simplified antimatter rocket shows how it produces thrust. Protons and antiprotons annihilate (far left) and produce positive, negative, and neutral pions. The neutral pions (bottom left) decay into gamma rays. The positive and negative pions travel more than 20 meters before they decay. Superconducting coils around the engine keep them confined to the rocket nozzle. They eventually decay into muons, neutrinos, and gamma rays. Before that happens, though, they are in the rocket engine long enough to provide thrust. That comes either from heating a working fluid, or from the kinetic energy from the particles themselves.

are forced to travel in tight circles around the magnetic field lines. Most of them stay inside the reaction chamber. Those that try to go toward the front of the engine are turned around by the converging magnetic field and sent back the other way. In a well-designed antimatter engine, most of the charged pions can be contained by the magnetic field in the engine and directed out the rear to provide thrust for the rocket.

Even the charged pions decay, but they decay into other energetic charged particles called mu-mesons or *muons*. (They also produce some neutrinos, the "ghost particles" of particle physics. Like photons, neutrinos have no mass. Unlike photons, neutrinos can travel through opaque matter. A thousand miles of solid lead is like a transparent window to neutrinos. However, neutrinos play no part in mirror matter propulsion. So consider this an interesting sideways glance at the truly strange nature of Nature.) The charged muons have a normal lifetime of 2.2 microseconds (2.2 millionths of a second), but because of their high velocity they live 6.2 microseconds. That's long enough to travel over a mile. This is certainly a great enough distance for them to make their way out of all but the longest rocket nozzles!

The charged muons finally decay into electrons or positrons, plus some more neutrinos. In principle, even more energy and propulsive thrust can be obtained from these high-speed charged electrons and positrons. It is only some time later, depending upon the density of the plasma that is generated, that the positrons and the electrons will annihilate to produce some high-energy gamma rays. In one sense, the end result of the annihilation of an antiproton with a proton is nothing but high-energy gamma rays and high-energy neutrinos. But if the propulsion engineers do their job properly, it is possible to convert a significant percentage of the initial annihilation energy into kinetic energy of the rocket vehicle by obtaining thrust from the intermediate high-energy charged massive particles. The conclusion is exciting: If milligram quantities of antiprotons can be made and stored, then they can be used as a highly efficient propulsion energy source. All of this is in line with presently known physical principles.

There's another way to use the charged pions from proton-antiproton annihilation: to extract energy from them by passing them through some gas or liquid such as water, methane, or liquid normal hydrogen. The gas or liquid will be used as the "working fluid" in the rocket engine. The electrical charge on the charged pions will interact with the electrons in orbit around the atoms in the gas or liquid. Very quickly, the charged pions will transfer their energy to the gas or liquid, heating it up. The heated "working fluid" in the reaction chamber of the rocket then passes through the nozzle of the rocket and is converted into a stream of hot gas flowing out in one direction, resulting in the rocket moving off in the opposite direction.

In this kind of antimatter rocket, between 30 and 50 percent of the annihilation energy from the antimatter fuel ends up as kinetic energy of the rocket exhaust. In a typical antiproton rocket design where the working fluid is hydrogen gas, 10 milligrams of antimatter (a single grain of antisalt) is equivalent in propulsion energy to 200 metric tons of chemical fuel.

Smashing Assumptions

The availability of antimatter as an energy source for space propulsion will revolutionize space travel. Many of the present assumptions implicit in the design of a space mission will no longer be valid. Mission designers will have to develop a new set of assumptions to replace them. The concepts of Δv and mass ratio play an important part in the present design of space missions. Assumptions about them will change.

Consider an unmanned mission to Saturn's rings. A robot space probe will actually take up residence inside Saturn's rings, right in the middle of the billions of particles that comprise that beautiful structure. The mission characteristic velocity includes (1) the startup velocity to get the vehicle up to speed, (2) the velocity changes needed to switch from the plane of Earth's orbit to the plane of Saturn's orbit (they're not the same), (3) the changes in velocity needed to to make sure the spacecraft crosses Saturn's orbit when Saturn is there, (4) the velocity change needed to slow the spacecraft down and match its speed with Saturn, (5) the velocity change needed to change the vehicle's orbit plane from the plane of Saturn's orbit to the plane of Saturn's rings (they're not the same, either), (6) the velocity change needed to drop into Saturn's gravity well, and (7) the velocity change to match the spacecraft's velocity with that of the particles in the ring. After the exploration has taken place and samples have been obtained, then all the maneuvers have to be repeated *in reverse*. All of those velocity changes, added up, equal the total characteristic mission velocity. Which is a measure of how difficult the mission is. The characteristic velocity of this Saturn ring rendezvous and sample return is a minimum of 48 km/s. That's ten times harder than going from low Earth orbit to geostationary Earth orbit.

Each chemical fuel has its own maximum exhaust velocity, depending basically on the amount of energy per pound in the fuel. The most energetic chemical fuel presently available is a mixture of liquid hydrogen and liquid oxygen. When burned under ideal conditions, its exhaust velocity is about 5 km/s. The Space Shuttle main engines, which use liquid hydrogen and oxygen, are remarkably efficient, achieving an actual exhaust velocity of 4.5 km/s, or 90 percent of the theoretical maximum.

Orbital velocity near the top of the atmosphere is 7.7 km/s, however, which is more than the exhaust velocity of the rocket engines. At first glance it might seem that the Space Shuttle can't reach orbital velocity. It does, of course, but at a price. The Shuttle uses up some of its fuel to lift off, and also to carry more fuel along to use later. After the rocket is moving, say at 1 km/s, the remaining fuel is also moving at a kilometer per second. The exhaust velocity is measured with respect to the vehicle, not to the ground, so the next batch of fuel coming out of the moving rocket will impart a greater change in velocity than did the first batch. That gets added on to the velocity the vehicle has already built up, making it possible for a rocket to slowly increase its velocity until it is going faster (with respect to the Earth) than the exhaust velocity of the fuel (with respect to the vehicle).

Table 7-4   Mass Ratios for Some "Impossible Missions"

Total mass ratio

The Missions	Mission V (km/s)	fuel: storable (Isp = 300 s (exhaust v = 3 km/s	LH2/02 500 s 5 km/s	nuclear 900 s) 9 km/s)
Reverse Orbit (X-wing fighter #1)	15.5	175	22	6
Double Reverse Orbit (X-wing fighter #2)	31	30,700	490	32
Solar Impact Mission (SIM)	35	117,000	1,100	49
Saturn Ring Rendezvous (SRR)	48	8,900,000	15,000	200

Some examples of mass ratios for certain "impossible missions" in the solar system. Only the single Reverse Orbit mission is theoretically possible with a form of "standard" space propulsion system available today (nuclear propulsion). All others are beyond the reach of current chemical or nuclear propulsion technologies. Taken from R. L. Forward, "Antiproton Annihilation Propulsion," Air Force Rocket Propulsion Laboratory, 1986.

The price the shuttle pays is that at the beginning of its flight it is using a lot of fuel to do nothing more than haul up more fuel. It's the "rocket equation" in action. And every chemical rocket must pay this price. The harder the mission, the larger the characteristic mission velocity, and the more fuel the vehicle has to have at takeoff for the same vehicle mass. The extra amount of fuel needed does not rise linearly with the increasing characteristic mission velocity; it rises exponentially.

What if the mission characteristic velocity is ten times the exhaust velocity, which is what's needed for the Saturn ring rendezvous mission? In that case, the mass of fuel needed is 22,000 times the mass of the empty vehicle. That's a mass ratio of 22,000! There is no way to build a bare fuel tank that would even hold 22,000 times as much fuel as it weighs, much less withstand launch stresses and deliver a payload. Obviously, missions that require such large mass ratios are impossible using chemical fuels. Some examples of impossible space missions are listed in Table 7-4. Mirror matter, however, smashes such assumptions. Mirror matter makes the impossible missions possible.

Missions Impossible

The exhaust velocity of chemical rockets is limited by the energy content of the fuel. This isn't the case with antimatter-powered rockets. The exhaust velocity can instead be tailored to match the characteristic velocity of the particular mission. That minimizes the mass ratio and also the cost of the space mission. Simple studies already done have shown that the reaction mass of an optimized antimatter rocket *never exceeds 4 tons.* The mass ratio is thus five tons initial mass (payload plus fuel) to one ton of final vehicle mass. In fact, reaction mass that minimizes total mission cost is typically 1.5 *tons for every ton of vehicle.* That's a mass ratio of 2.5. The amount of reaction mass needed in an antimatter rocket can nearly always be the same for any kind of mission. The same vehicle with the same load of reaction mass can be used for all missions. The only differences will be the amount of antimatter used and the type of antimatter engine used.

Instead of optimizing for minimum antimatter or minimum cost, suppose the mission were optimized for *minimum time* with reasonable cost restraints. Just the addition of a little more antimatter would decrease the trip times by huge amounts. With antimatter rockets, mission trajectories would not be long elliptical orbits, but nearly straight lines. A typical manned mission to Mars will no longer take years of time, but *months.* That in turn will result in significant savings in vehicle mass, life support requirements, and ground support costs. Research on antiproton annihilation propulsion technology admittedly carries a very high risk. The extremely high payoff in fast, efficient space propulsion, however, makes it worthwhile to expend a significant amount of research effort to determine the feasibility of the concept. A mirror matter space propulsion system may seem like an expensive proposition to us now. In the long run, though, it will save enormous amounts of money by using the energy of mirror matter annihilation to travel to the planets.

In addition, there seems to be a paucity of alternate propulsion concepts to solve the future propulsion needs of the Defense Department. Military and civilian planners have studied such "far out" concepts as advanced chemical propulsion, solar electric propulsion, solar thermal propulsion, nuclear electric propulsion, nuclear thermal propulsion, fusion propulsion, and even nuclear bomb pulse propulsion. None but the last (which is forbidden by space treaty and common sense) can achieve the results promised by antimatter propulsion. Rapid response time, high thrust combined with high exhaust velocity, short mission times, low system mass, high propulsion efficiency, and comparatively low cost—this is what space travel needs. Only mirror matter can deliver.

There is also the possibility that as research using antimatter uncovers more of the remaining mysteries of mirror matter, we may find much better ways to make mirror matter:

- It may be possible to find mirror matter pairs of exotic stable particles. These particles could be stored separately in containers of ordinary matter without annihilation on contact with the walls. Then, when mixed with their exotic antiparticles, they would completely convert into energy.
- It may also be possible someday to find some method of converting energy directly to antiprotons by a more subtle process than slamming a high-energy proton into a heavy metal target. The efficiency of making antiprotons could then rise from 5 percent to near 100 percent, and that would dramatically cut the cost of making mirror matter.

Turn Left at the Moon: V

The next "morning" the vacationers were taken out on a Dyson City shuttle to the Rockwell IT-2 Longliner operated by Mars Tours Unlimited. The Longliner was aptly named. It was five hundred meters in length. The main body was a long girder in the shape of a triangular truss structure, which connected the three main components together. At the front was the control center and passenger module. Near the center of the truss were the propellant tanks full of liquid methane. At the rear were the liquid drop radiators, the radiation shield, and the antimatter rocket engine.

Since the engines were off, the shuttlecraft took them near the rear of the Longliner first, to let the passengers see the marvel that had opened up the solar system to tourism. The Longliner's scotty was traveling out to the ship with them, and she gave them a detailed description of her pride and joy. Eric listened with an intense fascination that the scotty noticed, and mentally noted. *This kid's a comer,* she thought to herself.

"The antimatter rocket engine," she said aloud to the people seated with her on the shuttle, "is not much bigger than some of the chemical engines used on the big dumb boosters back in the 1990s. It's three meters in diameter inside the annihilation chamber, and the bell of the nozzle expands out to ten meters to get the maximum thrust out of the hot exhaust gases. In each second of operation, the engine uses three milligrams of antimatter to heat 1.4 kilos of propellant. The power level is 500 gigs—gigawatts, sorry—about 15 times the power level of the pioneering Saturn Five moon rockets."

The shuttlecraft moved to the other side of a thick metal disk three meters across. It was 15 centimeters thick at the center and tapered down to a few centimeters at the rim. Right in the back of the disk was a large rectangle the size of a kitchen refrigerator.

"The metal disk you can see is the radiation shield," said the scotty. "The rectangular box is the antimatter storage container. Antihydrogen pellets weighing about 10 micrograms each are extracted from the electrostatic storage container and sent down a vacuum line that runs through the shield. The pellets enter the center of the reaction chamber of the engine and annihilate with the carbon and hydrogen atoms in the methane. The reaction products are gamma rays and charged particles called pions. Most of the gammas get stopped in the walls of the reaction chamber, and they preheat the methane. Some of them get out of the engine, so we need that shadow shield to protect passengers and crew up front.

"The charged pions get trapped by the strong magnetic field bottle that's generated by the high-temp superconductor loops—you can see them wrapped around the pressure chamber and nozzle. The field also helps keep the hot propellant from getting too close to the walls and melting it.

"Oh, all the materials that make up the engine are specially chosen low-mass isotopes that don't produce long-lived radioactive byproducts if they get fissioned by gammas or pions. Within a day after shutting off the engine it's safe to work on. Thank heavens." Eric and several others chuckled.

The shuttle next passed by three long narrow troughs that stuck out 30 meters to each side of the truss backbone of the ship. Further along the truss were three 30-meter-long arrays of tiny spray nozzles.

"Ah," said the scotty. "The squirters and catchers for the liquid drop radiators. When the antimatter engine is on and we need to get rid of waste heat, the liquid drop radiators are turned on, too. Red hot drops of liquid metal come streaming out of the squirter nozzles and stream into space, where they cool off. The cooled droplets get caught by the collectors sticking out below, and the liquid metal is piped down to the engine compartments to cool them off. Then the hot liquid metal is returned to the squirters again."

Next they passed the large propellant tanks at the center of the Longliner. Three of them, each a squat cylinder 15 meters wide and three meters thick, lay stacked along the truss like beads on a stick.

"The tanks hold the reaction mass that's heated by the antimatter in the reaction chamber. Theoretically, you could use anything you wanted for propellant. I've been on ships that used water, that used hydrogen—they both work fine. But I think methane's best. It has four hydrogen atoms for each carbon, so it's almost as good as pure hydrogen. But it's more dense than liquid hydrogen, so the size of the tanks is more reasonable."

She paused a moment as the tanks glided by their view. "They hold 800 tons of methane. That's about four times the dry weight of the Longliner. If we were going to Saturn's rings instead of to Mars, we'd still only need 800 tons of propellant. We'd just take more antimatter to annihilate with it and so heat the propellant hotter.

"The reason for three tanks, by the way, is safety. If one of them is penetrated by a meteor and we lose propellant, we can make do with the contents of the other two. We just heat the stuff up more with a little more antimatter and get more total thrust with less propellant."

The shuttlecraft glided along the length of the Longliner, as the scotty continued talking and answering questions. They finally arrived at the passenger module, boarded, and found their rooms. Eric and his grandmother shared an outer stateroom with a heavy porthole. His parents had an inner cabin. Everyone spent most of the time on the upper deck with the large viewing window. There were also monitors for the remote cams that let them look aft, and for the long-distance cams that were coupled to powerful telescopes and gave close-up views of distant objects.

The Longliner's captain, a distinguished-looking man with wavy white hair and a handlebar moustache that impressed the dickens out of Eric, met them in the lounge on the upper deck and described the trip's details.

"We will first make a short hop to the Moon to pick up some more passengers and some supplies for the Mars base. We then turn left at the Moon and head straight for Mars. Now, in the old days, with limited rocket power, it was necessary to take a long roundabout elliptical orbit from one planet to another. You had to 'lead' the target planet by almost half an orbit. With the increased rocket power that antimatter technology has given us, we just aim for the target planet and drive straight to it. Mars is fairly close this time of the year, so we only have to travel about one astronomical unit, or AU. That's the distance between the Sun and the Earth, about 150 million kilometers.

"So that you will be comfortable, we will travel at a constant acceleration for half the journey, go into freefall for a few minutes while we flip the liner 180 degrees, and then decelerate constantly until we are in martian orbit. We will travel at one-sixth Earth gravity, one-sixth of a gee. That's one lunar gee, and about one-half a martian gee. Our total travel time to Mars will be seven days.

"I hope you enjoy the journey. If you have any questions, please don't hesitate to ask the first officer, or me if you can find me—and I hear our scotty likes to talk, too!"

Chapter Eight

Getting Down
to Earth

Mirror matter, as we've seen, is already being used right here on Earth by physicists exploring the microworld of quarks and leptons. Of course, there's nothing "practical" about slamming protons and antiprotons together at gigavolts of energy. The experiments at places like CERN, Fermilab, and IHEP are basic research. However, mirror matter *does* have practical uses, and right here on Earth, today. A medical imaging technique called PET, or *positron emission tomography*, is a standard procedure in hospitals around the world. It saves people's lives, and it uses antimatter—positrons.

Mirror matter will have practical uses in the future, as well. As we'll now see, some very exciting things can be done right here on Earth using extremely tiny amounts of mirror matter, amounts far less than that needed for rocket propulsion systems.

Mirror Matter—High Octane?

The U.S. Air Force has shown considerable interest in antimatter, and we'll look at that in detail in the next chapter. Of course, their interest makes sense: Mirror matter could become the ultimate source of fuel or energy for a space propulsion system. But what about propulsion *on Earth?* Can mirror matter be used to run airplanes, cars, trains, boats? Probably not, at least not in the next 50 years or so. One thing that we have certainly learned so far about mirror matter is that it is very expensive to produce. It will remain extremely expensive for some time to come, even with the development of factories expressly designed for its efficient production.

Mirror matter will be an economically competitive energy source for space propulsion, but not because it will be cheap. Rather, the energy available from matter/mirror matter annihilation is so enormous that even at a cost of $10 million per milligram, a mirror matter-fueled space transportation system will be less expensive overall than a chemically fueled space transportation system. The present American space transportation system is based on the Space Shuttle, its main tank booster, and its solid rocket boosters (Figure 8-1). The shuttles are expensive to construct and use because they must carry a lot of chemical fuel without weighing too much themselves. They expend enormous amounts of energy to get from Earth into orbit. Moving great masses of people and equipment from place to place in the solar system will also require very powerful propulsion systems. So will a military space cruiser/"X-Wing Fighter" capable of dogfighting maneuvers in orbit. Any energy source that can vastly outstrip what's already available and do so at competitive prices becomes a serious candidate for space propulsion use.

That's why mirror matter is an *unlikely* candidate for ground-based propulsion. There is simply no conceivable way that it will be possible to produce and handle mirror matter more cheaply and safely than, say,

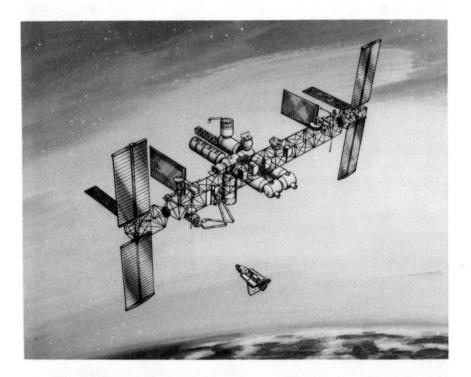

Figure 8-1   The space shuttle, shown here docking with the American space station planned for the 1990s, is the United States' only crewed space vehicle. Its main booster rocket uses liquid hydrogen and oxygen, the most powerful fuel easily available today. However, the shuttle still can not travel further than a few hundred miles from the Earth's surface. (*Courtesy McDonnell Douglas*)

gasoline or jet fuel. It will be the ongoing breakthroughs in warm and room-temperature superconductors that will more likely cause a revolution in transportation on Earth. Cheap and efficient magnetic-levitation commuter trains, transcontinental tunnel-trains using superconducting rails, and even cost-competitive electric cars using superconductor batteries are now distinct possibilities. Those developments could be upon us within the next ten years. The same superconducting technology will also make it easier to to make large amounts of mirror matter in specially designed factories, of course. But even with room-temperature superconductor technology, mirror matter will never be as cheap as the coming maglev trains or electric cars.

Then there's the "fear factor." Americans these days fear anything that has a connection to radioactivity and radiation. The inevitable byproduct of matter/mirror matter annihilation is radiation. There are easy ways to shield people from its effects, and certainly mirror matter-propelled space launches are not going to be a danger to the general public. Such launches could safely take place at Edwards Air Force Base, for example, or at the Atlantic coast of South America, where the European Space Agency launches its rockets. Space-based mirror matter propulsion, too, will be harmless to "flatlanders."

The same cannot be said with one hundred percent certainty about mirror matter-powered planes, trains, cars, or boats. Planes crash, trains collide, cars catch fire. No one, not even the most ardent mirror matter enthusiast, wants to see a fizzling mirror matter car engine in Times Square. We should not fall into the trap that so many nuclear energy enthusiasts tumbled into right after World War II. They thought nuclear power was a magic wand that would make everything possible. Then there were articles about nuclear-powered planes, trains, cars, boats, and possibly even wheelchairs. The military actually spent billions of dollars trying to build a nuclear-powered bomber. It never happened. The same problems with high-energy neutrons that grounded nuclear-fission rockets put an end to nuclear-fission-powered airplanes. Mirror matter is no magic wand, either. There are some tasks for which mirror matter is simply not appropriate or even useful.

Undersea Mirror Matter?

Unlike standard sources of energy for operating power plants, such as sunlight, coal, oil, or biomass, there are no "free" sources of mirror matter ready to be collected, mined, pumped, or harvested. Mirror matter is a synthetic fuel, and it must be made, one atom at a time. All the methods that we presently know of for making antiprotons require at least 250 times as much energy to make the antiproton than we will ever get out of it from its annihilation. For standard power plant uses, it would be better to use the energy directly for whatever purpose is desired, instead of going through the very inefficient intermediate process of producing antimatter. This low energy conversion efficiency would seem to preclude any use of mirror matter as a standard source of power for cities, homes, or industry on Earth.

However, that doesn't mean there would be no uses on Earth for mirror matter power plants. There are some situations on Earth where they would be useful—or perhaps we should not say "on Earth" but "under water." Is it possible that nuclear-powered submarines, for example, might become relics of the past, consigned to the history books along with sailing ships and plywood PT boats? A tungsten thermal core mirror matter power plant (or Augenstein power plant described in detail in the next chapter) might be just the thing for a submarine: It would weigh less, take up less room, and provide the submersible with enormous amounts of power. The reaction fluid for the power plant—water—would also be rather plentiful.

However, there is a problem with mirror-matter powered submarines. Nuclear reactors aboard submarines do not and cannot explode like nuclear weapons. There is very little risk of radioactive contamination of the sea should a nuclear-powered submarine sink—as several have. The world's oceans are very big with respect to the amount of contamination released. The mirror matter in a power plant would annihilate if it came in contact with normal matter like imploding seawater. That could cause a nuclear-type explosion, and the explosion could be pretty big. However, it would be nothing compared to even the smallest nuclear fission explosion. Nor would it produce any long-lasting radioactive debris. In any case, such an accident could be avoided using the same careful construction and safety features built into current fission reactors aboard present-day nuclear submarines. Although several Soviet and American nuclear submersibles have been lost at sea, there has never been an explosion from their reactors, nor even a major release of radioactive material into the oceans.

If military submarines could use mirror matter power plants, so could civilian research submersibles. Granted, they would be very expensive. However, similiar scientific efforts have a long history of support from governments and private grants. Such submersibles could be made as big and as strong as the researchers wanted; their power plants would still take them anywhere. That could easily include the bottom of the Marianas Trench, the deepest spot on the planet, for a long, leisurely exploratory cruise. After their week(s)-long expedition, the researchers would return in their comfortable submersible to their even-more comfortable undersea living quarters. Mirror matter power plants might be just the thing for an extensive undersea "village" for future oceanographers, subsea geologists, and ocean-bottom miners. Each power plant would be located in its own reinforced dome some distance from the main village. Power would be sent to the village through special room-temperature superconducting cables laid on the ocean floor. The pressure domes for the village and power plant would have been built out of specially reinforced materials by workers on the surface using mirror matter-powered undersea robots.

Mirror matter power plants might be very expensive to build, but that expense could be justified for special uses such as undersea submersibles or ocean-floor colonies. Indeed, mirror matter might be the key that opens the door not only to outer space, but also to the inner space of the world's oceans.

It is highly unlikely that mirror matter will be used for ground propulsion. It is somewhat unlikely, but possible, that it will find a niche on Earth as a specialized power source. There are, however, other uses for mirror matter right here on Earth that are quite likely, and that may be realized sooner than most people expect. Scientific, medical, metallurgical, engineering, and educational uses for mirror matter are all possible within the next 20 years. One way of looking at those uses is in terms of the amount of mirror matter needed for them. Table 8–1 gives an overview of the ways

Table 8-1   Using Mirror Matter: A Timeline

Time Frame	Scale (numbers of antiprotons, $\bar{p}$, needed)	Annihilation Energy (joules)
	FEMTOGRAM $(6 \times 10^8 \, \bar{p})$	0.18
Some research in progress	Low velocity nuclear physics Experimental non-destructive analysis of solids Vacuum measurement	
	PICOGRAM $(6 \times 10^{11} \, \bar{p})$	1.8×10^2
2–5 years circa $10M	Local 3D density imaging in solids Gravitational mass experiments High-density quark-gluon physics Trace 3D pathways in solid crystals Treat inoperable tumors, cauterize lesions Anneal micro-faults deep in crystals and metals	
	NANOGRAM $(6 \times 10^{14} \, \bar{p})$	1.8×10^5
5–10 years? circa $50M	Long-range interior imaging and analysis Small thruster experiments—10KW for 9 seconds Commercial NDE/NDA	
	MICROGRAM $(6 \times 10^{17} \, \bar{p})$	1.8×10^8
15 years? circa $100M	Medium thruster experiments (100 lb class) Small item industrial welding	
	MILLIGRAM $(6 \times 10^{20} \, \bar{p})$	1.8×10^{11}
20–50 years? $TBD	Large engine experiments (100,000 lb class) Small probe propulsion (100 lb class)	
	GRAM $(6 \times 10^{23} \, \bar{p})$	1.8×10^{14}
????	Space transportation (30 mg/30 tons payload to low Earth orbit)	

(Courtesy Maj. Gerald Nordley, USAF)

mirror matter could be used on Earth and in space, as arranged by the amount needed. The smallest quantity is a *femtogram*, one-quadrillionth of a gram (0.000000000000001 gram, or 0.00000000000000004 of an ounce), which adds up to about 60 billion antiprotons; the largest is in grams. The annihilation of a femtogram of antimatter will produce only 0.18 joules, not even hot enough to feel with your finger. A gram of mirror matter protons is 600 million quadrillion antiprotons. The annihilation of a gram releases 180 trillion joules, 30 times what's needed to put a 30-ton payload into low Earth orbit. Let's start, though, with a closer look at some down-to-earth uses of mirror matter.

Mirror Matter Physics

Physicists have been using mirror matter for decades as an essential part of their exploration of the subatomic microworld. That will continue. With sub-femtogram amounts of antiprotons, physicists are planning to make measurements of the fundamental properties of the antiproton itself. These efforts include the experiments by Gerald Gabrielse to measure the inertial mass of the antiproton (using only one antiproton). Here, too, we find the work being done by the Los Alamos National Laboratory and University of Pisa to measure the gravitational mass, using perhaps as many as a whole femtogram. At the Low Energy Antiproton Ring at CERN in Switzerland, physicists typically store a few tens of femtograms of antiprotons at a time in order to carry out very interesting work in low-energy nuclear physics. One particularly interesting area is the determination of the antiproton-proton and antiproton-heavy nucleus annihilation cross-sections: How many of the particles actually hit one another and annihilate? How efficient are these annihilation events at annihilation? Answers to the cross-section question will ultimately be needed for antimatter propulsion.

A picogram is one-trillionth of a gram. A picogram of antiprotons is about 600 billion (600,000,000,000) of them. The annihilation of a picogram of antiprotons would produce about 180 joules of energy, enough to light a 100-watt light bulb for 1.8 seconds. At Fermilab, the high-energy particle physicists want to save up an entire picogram of antiprotons after a few days of manufacturing them. Then they will put them in their Tevatron in order to collide them with normal matter protons at an energy of 2 TeV—two teravolts, or two *trillion* electron volts! In this way they will probe the interior of the proton and antiproton and perhaps uncover new breeds of quarks. Physicists could also use picograms of mirror matter to do high-density quark-gluon physics. These experiments could tell us if quarks are indeed the ultimate particles of matter, or if they, too, are made of still smaller entities.

Mirror Matter Medicine

Some of the most exciting uses for mirror matter will be in medicine. Not huge gram quantities, or even milligrams or micrograms, but as little as a femtogram of mirror matter is all that's needed (Figure 8–2).

Ted Kalogeropoulos, a physicist at Syracuse University in New York, and his colleague Levi Gray, have found that a femtogram beam of antiprotons can be used to create images of the interior of an object. That object could easily be a human body. They have conducted simulated experiments on a computer which have led them to believe that this new form of imaging will have several significant advantages over current methods of taking pictures of the inside of the human body, such as CAT scans. They have already given the process a name: *antiprotonic radiography,* or APR.

The CAT scan is probably the best-known method for making detailed pictures of the inside of the human body. CAT stands for *computer-aided tomography,* and tomography is a word that comes from the Greek words *tomos,* cut or slice, and *graphos,* picture. So tomography is literally "a

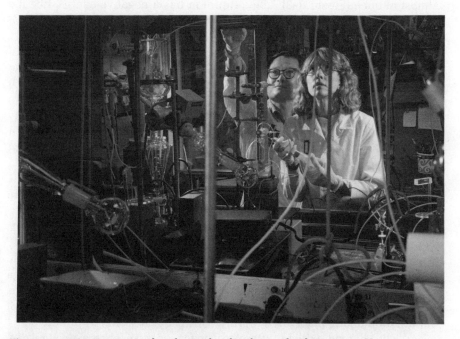

Figure 8–2 Antimatter is already used today for medical purposes. Here chemists incorporate a radioactive element into a compound injected into a person's brain. As the radiotracer decays, it emits positrons (antimatter electrons), which are detected by a special device called a *PET* (Positron-emission tomography) scanner. PET scanners use antimatter to locate small brain tumors and to study schizophrenia. *(Courtesy Brookhaven National Laboratory)*

picture of a slice" of something. In this case the slice is a cross-section of part of the body. A CAT scanner takes a series of X-ray pictures through one part of the body, and a powerful computer combines those pictures into a single image of a slice, or plane, cutting through the person's body.   CAT images are much more detailed than ordinary X-ray images, but there's a price paid for that detail. The person undergoes a relatively high exposure to X rays in order to make a cross-section image. Another shortcoming to CAT images is that they're not "real"—that is, the image that appears on the TV screen is not coming directly from the body. Rather, it is an artificial picture, constructed by a computer from many different X-ray images. Such artificial pictures could contain artifacts caused by the inherent limitations of the computational process.

Gray and Kalogeropoulos believe that APR could overcome both of these CAT scan shortcomings. APR would have other advantages of its own. Before we get into more details, though, let's look at how antiprotonic radiography would work.

When an antiproton penetrates some object, whether it be a copper target in a particle accelerator or a human body, it begins slowing down almost at once (Figure 8-3). The antiproton has a negative charge, like the electrons of the atoms in the target. As it passes near atoms in the object it ionizes them by repelling some of their electrons out of their orbital shells. In the process the antiproton ricochets off nearby atoms, loses its energy of motion (kinetic energy), and slows down. Finally, the antiproton comes to a complete halt. The place where it comes to a halt is highly predictable if the density distribution of the object is known. An antiproton with a given initial kinetic energy has a "range" in the target material that is inversely proportional to the density of the material. That means that an operator who knows the average density of an object (like the human body) can adjust the injection energy of the antiproton so that it will travel a selected distance into the object and come to a stop with an error of only a few millimeters (Figure 8-4). This allows antiprotons to be "placed" anywhere desired by aiming carefully and adjusting the injection energy properly.

While the antiproton is traveling through the object on its way to its stopping point, it will be moving by the atomic nuclei inside the object fairly rapidly. The high speed and the small diameter of the atomic nuclei inside an atom insure that most of the antiprotons reach the end of their range without ever hitting another nucleus and annihilating. Once the antiprotons have come to a stop, however, the story is different. The probability of annihilation taking place rises rapidly. This is because an antiproton, being negatively charged, will be attracted by the positive charge of a nearby atomic nucleus, and be drawn toward it. As it approaches, the atomic nucleus of the antiproton, being negatively charged, takes the place of one of the negatively charged electrons in orbit around the nucleus of the atom. It is like a "heavy" electron that weighs 1,836 times as much as a normal electron.

The antiproton starts out in a large orbit a good distance away from the

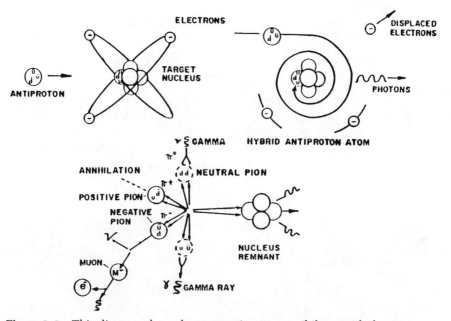

Figure 8-3 This diagram shows how an antiproton annihilates with the nucleus of an atom. As an antiproton travels through any piece of normal matter, it bounces off the electron "shells" of the atoms until it comes to a stop next to a normal atom. When the antiproton stops, the normal atom captures it. The antiproton displaces one of the electrons from its orbit. The antiproton acts like a super-heavy electron. However, the attraction between the positively charged protons in the nucleus and the negatively charged and heavy antiproton causes the antiproton to jump down from one electron level to the next. It quickly spirals into the atom's nucleus until the strong nuclear attraction between the two pulls the antiproton into the nucleus. When the antiproton spirals into the nucleus of an atom, the two annihilate. That causes the creation of gamma rays (top and bottom) and positive, negative, and neutral pions (left, top, and bottom). The pions eventually decay into other subatomic particles and finally into gamma rays. Often a remnant of the original nucleus is left and it too goes flying off (right). *(Courtesy Maj. Gerald Nordley, USAF)*

protons in the nucleus so the annihilation probability is still low. Under the attraction of the positively charged nucleus, the negatively charged antiproton jumps down from one orbit to the next. When an electron around that atom jumps down from a higher energy orbit to a lower energy orbit it emits a photon of a characteristic wavelength which we see as a color. In the same way, the jump of the antiproton also produces a photon of a characteristic "color," but it is a "color" we cannot see with our eyes, a high-energy X ray.

A chemist can take a small sample of material, burn it in a flame to excite the electrons in the atoms, and then measure the color of the photons that come off. Yellow indicates sodium, green indicates copper, and so on. In the same way, some time in the future, a physicist can send a beam of

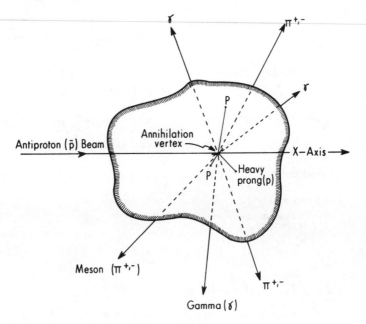

Figure 8-4   An antiproton with a given kinetic energy will travel a specific distance inside a target before it annihilates with an atom inside the target. If the energy of the antiproton and the density of the target are both known, one can predict where the antiproton will come to a stop and annihilate. On the average, three charged pions and four gamma rays are emitted. The average pion kinetic energy is 380 MeV. The pions escape with only small deflections. The gamma rays' average energy is 250 MeV. With no electrical charge, they escape the body undeflected. Special detectors can spot the escaping pions and gamma rays, trace their paths backward, and pinpoint the annihilation location. This technique can be used to build up a picture of the body's interior. In addition, the energy deposited by the antiproton beam can be used to destroy cancerous cells at the target point. *(Based on Fig. 1, Gray and Kalogeropolous, IEEE Transactions on Nuclear Science 1982, p. 1052.)*

antiprotons deep inside a sample of material, make the antiprotons come to a stop at some specific cubic millimeter deep inside, measure the "color" of the X rays that come out, and then tell us that there is some sodium, or aluminum, or uranium, or whatever, at that particular location inside the object.

As the antiproton drops down to the lowest energy orbits, it is getting closer and closer to its mirror twins inside the nucleus. Once it gets this close the *strong nuclear force* takes over. The strong nuclear force exerts its pull between nucleons when they are close enough to one another and it holds together the protons and neutrons in an atomic nucleus. This force now quickly pulls the antiproton out of its orbital shell and down into the nucleus. Result—annihilation.

The annihilation event usually occurs between the antiproton and a proton near the surface of the nucleus. It produces a spray of pions, gamma rays, and fragments of the nucleus. The neutral pions almost immediately turn into high-energy gamma rays that escape the body without interacting with the surrounding tissue. Most of the charged pions will also escape the body before they, too, decay. The nuclear fragments are relatively low-energy particles and don't travel very far in the body. In fact, they usually deposit most of their energy close to the annihilation point. This is important, and we'll see why later on.

What's important for antiprotonic imaging is what happens with the charged pions created in the annihilation. If the target (be it a thick casting or a human body) is not too large, the charged pions will almost always escape it before they themselves decay. The average free-flight path of a charged pion in water is about 30 centimeters or one foot. The average distance from the center of a human chest to the skin is only six inches. It is very likely, then, that charged pions coming from the center of a person's chest would escape. And they can be detected as they come flying out.

Since the gamma rays and charged pions leave in essentially a straight line from the annihilation point, their directions could be measured with particle detectors designed for that purpose. Once their directions of travel are computed, it would be a simple matter to determine the point in the body where the annihilation occurred. It will then be simple to produce, on a TV screen, a *direct image* of the annihilation point itself *immediately after the annihilation event.*

The pointing accuracy for such an imaging technique would be very high. Few pions will be deflected during their journey out of the body, and even fewer will be lost in the body by hitting another atom. Gray and Kalogeropoulos have computed that uncertainty in the location of the annihilation point would be about 2 mm, or three-hundredths of an inch. That's at least as good as the pointing accuracy of a CAT scanner. Methods exist that could improve APR pointing accuracy by a factor of two.

The "image" of a point inside an object is not very interesting, of course. What's interesting is an image made of a *series* of points. That could be easily done be using an antiproton beam to make a series of linear scans across a target, with each horizontal line just below the previous one. The position of each point on each line could be determined by a computer hooked to the detector array. This is very similar to the way the two Voyager space probes produced their spectacular images of Jupiter, Saturn, and Uranus. Each of those amazing pictures was actually a two-dimensional array of picture elements, or *pixels*. Each pixel represented one of 256 shades of gray. The spacecraft beamed the information back to Earth, where a computer reconstructed a picture made of 800 lines containing 800 pixels each, or 640,000 pixel points. An antiproton imager could do the same thing for the interior of a human body.

A picture made with a mirror matter beam might look like Figure 8–5. This is a simulated antiprotonic scan of two objects immersed in a bath

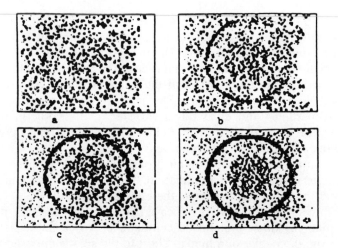

Figure 8-5   Simulated two-dimensional scan of a high-density cylinder contain-ing within it a lower-density rod. The two objects are in a pool of water. Using antiprotons in a scanning device, it may be possible to create images of flaws and defects inside solid objects. *(Based on Fig. 7, Gray and Kalogeropolous, IEEE Transactions on Nuclear Science 1982, p. 1056.)*

of water. The outer object is a hollow cylindrical ring with the density, thickness, and dimensions of a skull. Inside the ring is a rod with a density a bit less than that of the cylinder—we could pretend that it represents brain tissue. The (simulated) image in Figure 8-5a would be made using just five antiprotons for each line scanned (or five annihilation points per line), and measuring the paths and locations of the pions resulting from their annihilation. Figure 8-5b would result from using 10 antiprotons per line scanned; Figure 8-5c, 20 antiprotons; and Figure 8-5d, using 40 antiprotons per line scanned. A total of 150,000 (simulated) antiprotons were used to make the image in Figure 8-5d. If about 200 MeV of energy were released in the body for each annihilation, and that energy were deposited within a millimeter of the annihilation point, then the maximum radiation exposure for the picture in Figure 8-5d would be just 14-thousandths of a rad—much less than the radiation exposure from an average X ray.

The images in Figure 8-5 are of two objects in water. That's significant for imaging the interior of the human body, which itself is not much more dense than water. The simulation was done by pretending that the energy of the mirror matter beam was adjusted to deposit antiprotons ten centimeters deep into water before they would come to a halt.

A mirror matter beam could easily revolutionize medical imaging. It could also have another medical use—treating inoperable cancer tumors. The same beam that could image the tumor could also zap it; by increas-ing the number of antiprotons beamed at some particular point, doctors could deposit enough energy right in the tumor to destroy it. Similar methods would allow physicians to cauterize inoperable lesions deep inside a person's body. This kind of medical technology using mirror matter beams might be particularly useful on infants and small children. The energy in the nuclear fragments would be deposited right at the point of annihilation—the tumor or lesion—and thus perform the healing task. However, the dangerous pions and gamma rays would pass completely out of the children's tiny bodies before having a chance to hit any tissue and cause damage.

Carving Crystals

Picogram quantities of mirror matter will also have practical uses. Since slowly moving charged antiprotons can be controlled and directed by electrical and magnetic fields, researchers could direct them in complex paths to carve out three-dimensional pathways in space-grown crystals. It would open up a whole new era of engineering—*nano-engineering*. We would have radios the size of a molecule, the circuitry of a Cray supercomputer compressed into an area the size of a sugar cube. If self-aware intelligence is somehow connected to the density of neuronal pathways, then nano-engineering could lead to a sentient computer. Perhaps Isaac Asimov's fictional robots with positronic pathways in their brains will someday be real.

And one may hope that Asimov's Three Laws of Robotics will be positronically carved into their brains!

Analyzers and Sensors

So far we've looked at some of the things that could be done with femtograms and picograms of mirror matter. What about nanograms of the stuff, billionths of a gram?

Suppose we wanted to make an image of the interior of a solid casting of metal rather than a human body. The density of metal is much greater than that of water, but the energy of the mirror matter beam could be readjusted to compensate for that. For a moment, let us imagine that we want to examine the interior of a solid plate of titanium. Water, by definition, has a density of one gram per cubic centimeter, or 1.0. Steel has a density of 7.7 and titanium has a density of 4.5. The mirror matter beam's energy is "tuned" to penetrate into the titanium plate, and what we see is something like Figure 8–5d. However, the objects are not a skull and brain tissue, but several flaws inside the titanium plate. In the same way the antiproton beam could make an image of the interior of a human body, it can detect flaws in a solid block of metal.

However, we can do even better than that. By increasing the number of antiprotons deposited at a point per second, we can actually anneal the interior flaws we have detected. Within moments, a nanogram of mirror matter would heat up the local flawed region and eliminate it.

Still another use for tiny amounts of mirror matter would be the accurate X-ray analysis of unknown substances. As we mentioned earlier, physicists have learned something interesting about the X rays released just before an antiproton annihilates a proton or neutron in the nucleus of an atom. The X-ray pattern released just before annihilation is different for different elements, something like a fingerprint. Once we have portable containers for small numbers of antiprotons, we can extract them a few at a time and inject them into the interior of some unknown substance. Each burst of X rays coming from the sample will be characteristic of a particular element. The X-ray patterns from just 10,000 antiprotons will help to analyze the composition of a sample with an uncertainty of one percent.

Having hundreds of trillions of antiprotons available, however, may make possible a science fictional form of analysis: a new era of remote sensing devices. Picograms of mirror matter might enable us to peer into the interior of a human body with little radiation damage, or "see" the inside of a solid block of titanium. But that kind of imaging would have to be done with machines right next to the object being imaged. With nanogram quantities of mirror matter, however, it might be possible to develop long-range imaging capabilities. Such a beam would also be able to look into the interior of solid objects, and produce enough information for computers and humans to analyze the nature of the object's interior. But the interior imaging could be done from far away, perhaps hundreds of meters distant. This is not exactly the same as the "sensor beams" of Star Trek, but it is a step in that direction.

Mirror matter remote interior sensors would have many uses in industry and construction. For example, safety inspectors could check old buildings for hidden weaknesses in their steel girders. They could do the same on new buildings under construction, and do it quickly and efficiently with remote interior sensors.

Mirror Matter Goes to College

Nanograms of mirror matter could also be taken to school. A nanogram is one billionth of a gram. A nanogram of antimatter could contain 6×10^{14} antiprotons. The annihilation energy from an entire nanogram of antimatter should be treated with some respect, since it contains 180 kilojoules of energy. That sounds like a lot, but it will run a 100-watt light bulb for only a half an hour. The ability to store nanograms of mirror matter in compact magnetic traps for long periods of time could revolutionize university science education.

Imagine a lower-division university physics class that's studying the nature of matter and energy. They'll still have textbooks to read and homework to do. The students will also have lab sessions—but their physics lab will have a rather special piece of equipment. Off in a corner of the room, plugged into a wall outlet, will be a canister about the size of a street-corner mailbox. It will be a special kind of thermos bottle. A combination of high- and low-temperature superconducting magnets and a magnetic refrigerator will keep the inside walls of the antiproton "bottle" at a temperature near absolute zero. There will be a near-perfect vacuum in the bottle, since any residual air would have been frozen out on the walls. At the center of the bottle, suspended in that hard vacuum by powerful magnetic fields created by the warm-temperature superconducting magnets, will be a rotating cloud of a nanogram of mirror matter protons. That's a lot of mirror matter for a college lab. It would be more than enough to allow university physics students the chance to actually watch a particle of mirror matter annihilate a particle of matter, learning firsthand (and quite safely, we would add) the truth of $E=mc^2$.

Microgram Technology

The next step above nanograms is micrograms, millionths of a gram of mirror matter. It takes 6×10^{17} antiprotons to mass a microgram. The annihilation energy obtainable from a microgram of antiprotons is technologically significant: 180 megajoules, the energy equivalent of 20 kilograms of chemical fuel (5 gallons of gas). We are perhaps just five to ten years from nanograms of mirror matter to work with. Micrograms of the stuff is further in the future, perhaps fifteen years distant.

To store micrograms of antimatter, we will probably have to develop a method of producing antihydrogen ice instead of storing our antimatter as antiproton ions. But with micrograms of antimatter readily available, antiprotonic radiography will no longer be confined to major hospitals and medical centers. Micrograms of antiprotons will be made and stored by chemical supply companies, who will make nanogram amounts economically available to community colleges and high schools, and not just universities. More cities and municipalities will be able to afford mirror matter remote interior imagers.

Beyond Micrograms

Micrograms add up to milligrams, thousandths of a gram of mirror matter, or 6×10^{20} (600 million trillion) antiprotons. We can now almost see the tiny milligram speck of antihydrogen, bloated because of its small density of 0.08 grams per cubic centimeter to the size of a grain of salt. That tiny speck has the ability to release the energy content of 20 tons of rocket fuel. All of the down-to-earth applications of mirror matter mentioned so far needed far less than milligrams of mirror matter for their feasibility. Milligrams of the stuff, economically available, will simply make these applications more likely and less expensive. Augenstein tungsten thermal-core mirror-matter power plants may become more common in submarines. They might even be considered for undersea colonies of researchers or ocean-bottom miners.

Finally, with the availability of microgram and milligram amounts of mirror matter, it becomes possible to begin building and testing rocket thrusters using mirror matter. Small thruster experiments—say, ten pounds of thrust for a few seconds at a time—will be possible with nanograms of mirror matter. We can carry out medium thruster experiments, in the hundred pound-thrust range, with microgram amounts of mirror matter. Milligrams of mirror matter make possible large engine experiments. Such engines would produce tens to hundreds of thousands of pounds of thrust.

At this point we would be talking about real space propulsion systems, even including Earth-to-orbit launches. Milligrams of mirror matter could also be used to power propulsion systems for small unmanned space probes. At this point, we are not talking "down-to-earth" any more, but "out into space."

Finally, perhaps fifty years or more in the future, we will have the ability to produce grams of mirror matter per year, and store it indefinitely and cheaply. At that point mirror matter space transportation becomes an economic reality. An Earth-to-orbit launch system using 30 mg of mirror matter will put 30 tons of payload into low Earth orbit for less than ten percent the cost of a Shuttle launch. Beyond that, grams of mirror matter make it possible to go literally anywhere in the solar system at costs per passenger not too different from a luxury ocean cruise today. Not everyone will be able to afford a luxury vacation trip to Mars—but then, not everyone today can afford a luxury vacation cruise to Australia.

On the other hand, there is simply no way to predict how a technology as powerful and far-reaching as mirror matter technology will change society. One thing we can predict, though, with complete confidence: When mirror matter is available in gram quantities on an ongoing basis, society will be changed. Let's not make the mistakes of some past futurists, who envisioned a twentieth century with steam-powered cars and elaborate wrought-iron cities, clean air and water, and absolutely no hint of thermonuclear weapons, mass starvation, and state-sponsored terror and torture. We can instead learn from our history, and from the successes of other past futurists. Jules Verne's *20,000 Leagues under the Sea* is a study in the potential in science for both good and evil. H.G. Wells's cautionary vision of the future in *The Time Machine* foreshadowed the horror of weapons of mass destruction. George Orwell saw a world frozen into a perpetual state of war in *1984*, in which totalitarian states allowed nothing new—not even love. A society with mirror matter technology will still be a human one, and human nature is not likely to change much in the next 20 to 50 years. People will still oppress other people, and others will protest that oppression and try to remedy it. Some people will still be poor and others starving, to the shame and dishonor of those who are well off and well fed.

And yet... and yet... Global communication is making the starvation and torture of millions vividly real to the billions. The standard of living is rising not just in the developed countries, but in many developing countries as well. Technologies with roots in the beginning of this century are partly responsible for the improvements in the lives of humans everywhere. It is reasonable to think that a technology with roots in the latter years of this century—mirror matter technology—could help improve things even more.

Perhaps not every high school in the developing countries will have a canister of mirror matter in the high school physics lab. But quite a few of them most certainly will. And perhaps not everyone in the world will be able to afford a luxury vacation to Mars aboard a mirror matter-powered space liner. But quite a few humans will be able to get away to the Moon for a weekend hike in the lunar Apennines and a visit to the Tranquillity Monument.

Most importantly, our vision will change. Perhaps for the first time, it will become binocular. With mirror matter technology providing the real possibility of nearly limitless energy, we will finally begin looking both inward at our own planet, finding ways to change conditions for the better—and outward, at space, the final frontier, the frontier that has no ending, a frontier worthy of our limitless drive to explore and to grow in knowledge.

Turn Left at the Moon: VI

E ric looked out the big viewing window in the lounge. The Mars Longliner was in high orbit above the Moon. Jim was especially interested in the lunar stop. CERN was building its next big particle accelerator on the Moon. The machines now being designed were so gargantuan they could not even fit in the continent of Europe. The ring would fit entirely around the Moon, hopping from mountaintop to mountaintop where possible, and from tower to tower over the lunar maria. This machine was called LUPEC, for LUnar Positron-Electron Collider. The Grand Old Man of particle physics, Carlo Rubbia, was the titular head of the project. After several years living in Dyson City's low-gee section, he had moved to CERN's lunar base seven years earlier, still alive and kicking and (according to rumor) generally having a very good time outliving everyone else of his generation.

LUPEC would be a marvel. A beam of electrons would circle the Moon in one direction while a beam of positrons circled in the other direction. At six different mountaintops spaced around the Moon, the two high-energy beams would intersect to produce particle-antiparticle collisions approaching a thousand teravolts of energy. LUPEC was to be a peta-electron-volt machine. (Japanese physicists referred to it in conversation as "PETAC," an allusion to their own TERAC machine on Hawaii).

LUPEC would be cheaper to build than the previous smaller machines on Earth. There was no need to dig an underground tunnel to provide a shield from the stray synchrotron radiation generated by the high-speed particles. There was certainly no need to install a vacuum pipe; the Moon still had all the vacuum anyone could want. (Ginny grumbled at times about lunar air pollution, but no one paid attention.)

Jim pointed out the towers and mountaintop construction sites to Eric as the Longliner passed over them. They swung around the Moon and passed over the backside. He showed Eric the location of a large installation that the LUPEC track passed near.

"That's the first large antimatter factory built off-planet," said Jim. "The Russians built it here in 2010, powering it with large nuclear reactors instead of solar cells. It was designed to produce about ten kilograms of antimatter a year."

"Why the backside of the Moon?" Eric asked.

"They didn't feel very confident about the safety of their antimatter storage systems. They're still suffering from Chernobyl Syndrome, I think."

Eric watched the lunar scenery roll by, and eventually dozed off in freefall. Jim strapped him in a chair in the lounge.

The klaxon announcing the starting of the engines for their journey to Mars woke Eric up.

"Huh?" he struggled a moment, confused. "Hey, are we there yet?"

The weeklong journey to Mars was not entirely uneventful. True, there were no meteor strikes, no solar storms, no space pirates. And no girls Eric's age. (For that last fact, Eric wasn't sure if he was happy or disappointed.)

On the second day, though, they passed close enough to an incoming comet that they could observe it with one of the telecams on the monitors. The ship's scotty was controlling the cam and offered some comments.

"It hasn't developed much of a tail yet, as you can see," she said, "and probably won't. This is an old one, and it doesn't come too close to the Sun. Comets like this one are where

we get the methane and other hydrocarbons we use for propellant in our antimatter ships. Underneath the dark carbon dust surface is a dirty snowball of frozen methane, water, carbon dioxide, and ammonia. We save the nitrogen for fertilizer and convert the rest into hydrocarbons and water."

Early on the fourth day came the turnaround. The captain had them go to their rooms and strap in for the maneuver. Kirsten had given Eric a spacesickness pill at breakfast, but he had dropped it while rolling it across the table. When turnaround came, Eric's breakfast turned around, too—all over the stateroom. Ginny caught most of it in midair with a towel before ship deceleration began and plastered it to the floor.

Later that day the captain made an unexpected announcement. "If you come up to the main viewing lounge, you may be able to see one of the fastest objects in the solar system."

When the passengers had all crowded in, the captain said, "There's a Federal Express overnight delivery rocket from Earth to Mars catching up with us. It's just made turnover and its nozzle is pointed in our general direction. We've made turnover, too, and our nose is pointed towards Earth, so we should be able to see the flare of the exhaust from the front view windows."

He dimmed the lounge lights, then shut them off. They all looked out at the bright star-like spot moving visibly toward them. The scotty had pointed a telecam at the courier, but all that showed on the screen was a glare of antimatter-powered exhaust.

"What's it carrying?" someone asked.

"There was an electrical fire in the

control room of the antimatter power plant at Tharsis Base," said the captain. "The express rocket is bringing in a new control computer. Those little rockets are really engineering marvels. They put this liner to shame. They accelerate at ten gees and reach more than one percent of light speed at the halfway point to Mars. Then they flip over and decelerate at ten gees, and get to Mars in about 22 hours. A true interplanetary overnight express. It's a wonder to me that engines survive the heat, stress, radiation, and acceleration."

"It isn't to me," said Kirsten.

The captain looked at her, surprised.

She shrugged. "It's one of my rocket engines," she said. "My plant at Turin builds the engines for Federal Express. And my team scans 'em all before they go out. Only the best."

Eric grinned.

On the seventh day Eric woke up early. "Are we there yet?" he mumbled to Ginny.

"Almost," she replied. "I can see Mars at the bottom of our viewport. Get dressed and we can go out and watch them dock at Phobos.

"Here." She handed Eric a pill and a squeezebottle of water from the washstand, and waited until she saw him swallow it.

After breakfast the klaxon sounded, and the antimatter engines shut down. The Longliner had matched orbits with Phobos. The landing crew at Singer Station brought out a flexible passageway to connect the ship to the tiny moon.

They had half a day on Phobos before they went down to the martian surface. A guide took them on a tour of some sites near the Station. Then

they returned to the Longliner for dinner and another night's sleep.

Early the next day they boarded the Mars aerospace plane that would take them to the surface of the Red Planet. The McDonnell-Douglas AS-3M Marsclimber was a modified version of the Spaceclimber that had taken them from Palmdale to L-4. They drifted away from Singer Station using chemical rockets; then, as soon as they were a safe distance away, the antimatter-augmented rocket engines burst forth in a flaming plasma jet. The Marsclimber fell out of orbit and dropped toward the surface below. Since this was a vacation tour, the pilot slotted them into a low martian orbit for a few turns around the planet, and rolled the ship over so the passengers could get a good view. She gave them a guided tour of the surface for three orbits. The southern hemisphere was in winter, and the south polar icecap was fully developed.

Then the pilot fired up the engines again and dropped them out of low martian orbit for the spiral down to the surface. The extra large air intakes began to suck in the tenuous martian atmosphere. The engines switched to airbreathing mode. With no free oxygen for combustion in the 95 percent carbon dioxide martian air, the Marsclimber's supply of antimatter provided all the energy to heat the air for thrust.

The aerospace plane screamed in over the Solis Plain and landed on a long runway at Tharsis Base in the shadow of Mount Pavonis. A crawler came out to the plane, mated with the airlock, and took the tour group to the Penelope Hotel in the sprawling town of Burroughs outside the Base.

Burroughs was completely covered with a geodesic dome, since the martian atmosphere is only one percent as thick as that of Earth.

Eric's father stretched in the small hotel room. "Ahhhh. It sure feels good to have some weight back after all those days at low gee. Thirty-eight percent of Earth gravity is just fine.

You're bouncy but your feet stay on the ground."

"I liked the ship's gravity better," said Eric. "That felt like having seven league boots on. Here I'm feeling twice as heavy."

His grandmother laughed. "You are!"

Chapter Nine

To Provide for the Common Defense

From Magic to Matter-of-Fact

From the underground tunnels of the particle physics laboratories around the world, there is emerging a new energy source that will revolutionize the space operations of the Department of Defense and radically cut the long-range costs of the Strategic Defense Initiative. Huge multistage launch vehicles with strap-on boosters that are used once and then thrown away will become obsolete. The new energy source we speak of will power single-stage-to-orbit aerospace planes that will take off from a runway, deliver payloads to any orbit between Earth and the Moon, and return to base to be used again. This same energy source can then supply those payloads with nearly unlimited amounts of power, whether it be continuous (for surveillance) or in sudden bursts (for weapons). When the need is urgent or danger nears, this seemingly magical energy source will rapidly shift the payload to any desired new orbit without using large amounts of propellant. America's space forces will be able to carry out missions that are now impossible, while doing at higher speed and lower cost those missions that are already possible.

This new, lightweight, compact, high-efficiency energy source is stored mirror matter.

Energy from Mirror Matter

But isn't a mirror matter rocket only science fiction? Not at all. Using mirror matter as an energy source is simple. Suppose a mirror particle comes into close contact with a normal particle. The various properties that make up the two types of particles, such as charge and spin, cancel each other out. All that is left is the energy that used to be stored as mass in the two particles. As we learned back in Chapter 2, this equivalence of mass and energy is one of the revolutionary aspects of Albert Einstein's 1905 Theory of Special Relativity. In a very poetic sense, mass is "frozen energy." Energy is "evaporated mass." This complete conversion of mass to energy makes mirror matter a highly efficient, compact, lightweight source of energy. In fact, since the normal matter is easy to come by, one could say that mirror matter is the only energy source that is 200 percent efficient!

The major Air Force need is actually for lower cost access to space. Air Force satellites provide surveillance, warning, communications, weather, and space environment information to the United States (Figure 9–1). But it

Figure 9-1   The space shuttle is shown in this artist's illustration servicing a satellite that is part of the U.S. Strategic Defense Initiative. Note the space-suited figures for the scale of the satellite. A new generation of mirror matter-powered space planes would make this type of SDI mission much easier and less expensive in the long run. Currently, however, the Defense Department has no plans to use mirror matter as part of the SDI. (*Courtesy Lockheed Missiles & Space Company, Inc.*)

costs about $4,000 for every kilogram that gets into orbit. To save money, satellites are constructed to be as small and as lightweight as possible and still do their jobs. But that often means they're *too* small to do their jobs as well as they could. That's why the Air Force is working hard to develop new expendable rocket launchers. They will be cheaper to use and operate than the Space Shuttle. The planned "heavy lift launch vehicles" will also be able to put into orbit satellites that are much larger than those the Shuttle can handle (Figure 9-2).

However, a major propulsion breakthrough would change the rules of the game dramatically. Satellites could carry bigger antennas, larger telescopes, and more powerful communications equipment. They could be placed in orbits that would be out of the range of antisatellite weapons. In the future, military space personnel and systems may support space stations and even lunar bases in the same way they support polar operations today. They'll need to travel rapidly between the Earth and Moon (in *cislunar space*, as the astronomers would say). The answer? Mirror matter. Rocket engines powered by milligrams of mirror matter will enable the military to carry out

Figure 9-2   This partially reusable heavy lift launch vehicle is one design for possible use in the 1990s. The rocket's flyback booster would use liquid methane as fuel. *(Courtesy Martin Marietta)*

missions that simply cannot be done with chemical rockets or that would take too long using nuclear electric propulsion. More than half a century after the first dreams of such things, antimatter rockets will finally become a reality.

Space Launch!

The ARIES office of the Air Force Astronautics Laboratory in California is located at Edwards Air Force Base, in the high desert not too far from Palmdale, California. It's a desert scene that visitors to Edwards see: sand, scrubland, joshua trees. In the summer it gets beastly hot; but on January mornings the air carries a chill, and sometimes frost covers the ground. ARIES stands for Applied Research In Energy Storage. The office has as one of its two major goals the exploration of the feasibility of and possible military uses for antimatter propulsion.

It's the feeling of some people in the ARIES office that the best case for Air Force use of mirror matter propulsion is Earth-to-orbit launches. Edwards could become America's third major spaceport if that happened. Earth-to-orbit mirror matter propulsion could even be relatively near-term, say in the next ten to fifteen years. The reason is that it's possible to build a working antimatter engine with relatively simple Augenstein technology. "Augenstein" is Bruno Augenstein of the Rand Corporation in Santa Monica, California. He has devised a mirror matter space propulsion engine that uses a large block of porous tungsten metal (Figure 9–3). The annihilation

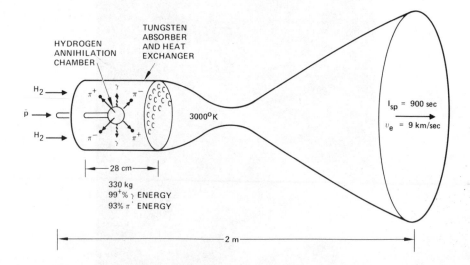

Figure 9-3 An Augenstein-type antimatter engine is a first-generation form of mirror matter space propulsion system, devised by RAND Corp. scientist Bruno Augenstein. An Augenstein mirror matter engine would have uses not only in space ships, but also on Earth.

of antiprotons and normal matter takes place in a hollow at the center of the block. Energetic subatomic particles and gamma rays from the annihilation heat the tungsten block. That heat in turn energizes the rocket propellant that is passing around and through the porous tungsten block.

An Augenstein engine could have an exhaust velocity of 9,000 to 10,000 meters per second. With that sort of an engine (Figure 9-4), in theory a

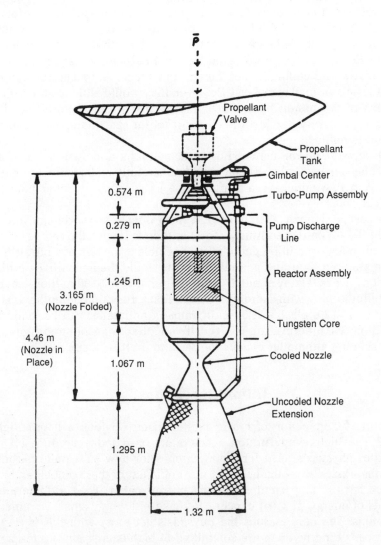

Figure 9-4   This more detailed drawing of an Augenstein rocket uses the basic design of the NERVA nuclear rocket tested in the 1960s. The nuclear reactor is replaced by the tungsten-bed Augenstein antimatter reactor module. (*Courtesy Dr. Steven Howe, Los Alamos National Laboratory*)

vehicle could be designed the size of an airliner that would carry the same payload as the Space Shuttle, and be able to fly itself into orbit. And Edwards Air Force Base *would* be ideal for a mirror matter rocket launch site: in the desert, far from large inhabited cities but with plenty of aerospace support facilities.

The same thing could be done with a nuclear reactor as the power source, but it would produce copious amounts of radiation. So would a mirror matter engine, but in the case of a nuclear reactor most of the radiation is in the form of energetic neutrons, which pose a serious hazard to the rocket's human crew and to ground personnel. That means lots of heavy shielding, and the advantages of high fuel economy are seriously compromised. Fully shielded nuclear-powered Earth-to-orbit rockets are possible. Researchers at Brookhaven National Laboratory have produced a design that would perform better than current chemical rockets (though not as well as mirror matter rockets). However, they would still face the serious problems of the fission by-products, radioactive waste disposal, possible accidents—and very quickly a nuclear rocket for space launches becomes very questionable.

Not so with mirror matter. Energetic neutrons and fission waste are not problems necessarily associated with a carefully engineered antimatter rocket. And it would have a performance as good as or better than the nuclear rocket.

The cost per launch is attractive, too. According to Major Gerald Nordley of the ARIES office, it costs roughly $100 million to launch the shuttle, with a weight of some five million pounds. If the vehicle is the size of a big airliner, say five hundred thousand pounds, the costs would theoretically be further cut. So now we're talking about $10 million rather than $100 million, leaving $90 million to play with. Nordley's calculations show that 30 milligrams of antimatter will do all this. If 30 milligrams of antimatter can be made for $30 million, $60 million a trip is saved. It's an attractive *economic* argument for considering antimatter rockets for Earth-to-orbit launch.

Antimatter Bombs?

One military application of mirror matter seems obvious: a bomb. Right? Wrong! The high-energy proton accelerators used today produce all the antimatter physicists need for their searches for new subatomic particles and Nobel Prizes. Even the best and most efficient of these machines, however, are terribly *inefficient* at making mirror matter. They use enormous amounts of energy, at great expense to make piddlingly small amounts of antiprotons. The only reasons the physicists get away with this is (1) the governments funding them are committed to high-energy physics research, and (2) it's the only way the scientists currently have of making mirror matter. The present low production efficiency of antiprotons has resulted in a high cost estimate for antimatter. And this has one major implication for military planners and would-be "antimatter terrorists": It makes no sense whatsoever to use mirror matter to make an "antimatter bomb."

The cost of the mirror matter is always likely to be much greater than the cost of uranium or lithium hydride, which are used to make fission and fusion bombs. These nuclear materials are relatively cheap, easy to manufacture, and especially easy to store as bombs for long periods of time without deteriorating in effectiveness. And they make very large bangs (many megatons, the equivalent of millions of tons of TNT, not measly kilotons), for very few bucks. A nuclear bomb makes good military sense since it can destroy targets that cost 1,000 to 100,000 times as much as the bomb costs.

Now, what about an antimatter bomb? Well, to make a single, small antimatter bomb with the energy release of the Hiroshima nuclear bomb (20,000 tons of TNT or 20 kilotons) would require a gram of mirror matter. That's a hundred times more mirror matter than needed for an antimatter rocket. A gram of mirror matter would cost *at least* $10 billion, or one percent of the total 1987 United States national budget of about $1,000 billion. All this, just to buy one small bomb? Any country trying to make antimatter bombs would go broke in the attempt. To make an antimatter bomb is economically absurd.

There's also a scientific reason for not making antimatter bombs. The penetrating power of the gamma rays and pion particles that come out of the matter-antimatter annihilation process is very high—so high, in fact, that the energy released in those high energy particles is spread out over a large volume of material. An antimatter bomb will not go "bang"; the heated material will go "poof" and make a poor explosion. Not only do antimatter bombs cost more than the objects they can destroy, they don't even work well! It's not much bang for an awful lot of bucks.

An Antimatter Weapon

There is one weapons application for antimatter that was considered briefly during the formative stages of the SDI: A neutral particle beam weapon. This would be a powerful particle accelerator, a compact version of the particle accelerators we've already discussed that accelerate protons to make antiprotons. The particles that would be accelerated in these proposed machines would not be protons, but negative hydrogen ions, or protons with two electrons in orbit around them. Since the negative hydrogen ion is charged, the accelerator will speed it up until each ion has a great deal of energy. The high-speed ions are then aimed at the distant target by an electromagnetic aiming lens.

If the negative hydrogen ions were left charged—or any other charged particles, such as antiprotons, were used as the "bullets" in this particle "gun" weapon—they would never get to the target. The earth has a magnetic field that is weak, but still strong enough to bend the paths of the high-speed charged particles. The charge on the ions would interact with the magnetic field. The ions would have a tendency to spiral around the magnetic field lines. They'd either head out into space or spiral into one of the Earth's magnetic poles. In order for a particle beam to go in a straight line from the aiming lens of the weapon to the target, it must not have a charge. This is easily arranged by having the high-speed negative hydrogen ions pass through a small "stripping chamber" of gas. The extra electron on the hydrogen ion is stripped off by a collision with the electrons around the gas atoms in the stripping chamber. That turns the negative hydrogen ion into a plain old neutral hydrogen atom, with one proton orbited by one electron. The now-neutral hydrogen atom speeds on its way straight to the target at close to the speed of light.

The neutral hydrogen atoms do not need to travel at the speed of light in order to get to the target on time. A tenth the speed of light will get the particles to a target 3,000 kilometers (1,800 miles) away in a tenth of a second. However, there is another reason they must travel near light speed. In order to do some damage to the target when it gets there, each hydrogen atom must be accelerated up to high energy, typically 2 GeV. And that means that the particle has to be accelerated to 94 percent of the speed of light. This will insure that when the hydrogen atoms get there, there is enough kinetic energy in each atom to generate a half-dozen or so pions, lots of electrons and positrons, and oodles of high-energy highly-penetrating gamma rays in the skin of the target. Those charged particles and gamma rays can easily penetrate to the interior of the target and fry the electronics so the guidance will not guide, the sensors will not sense, and the trigger will not trigger.

To give each neutral particle in the neutral particle beam weapon the energy needed to make sure the collision-created particles and gamma rays have enough energy to penetrate through any conceivable shield, the accelerator must be long and powerful. It will need a very large electrical power supply to operate it. The accelerator and the power supply will have to be large and massive, and it will cost a lot to put them into orbit.

However, there is an alternate method of creating a neutral particle beam weapon that will hit with the punch of a 2 GeV weapon, but can be operated for the mass, size, and cost of a 0.2 GeV weapon: That's to use positive *antihydrogen* ions as the bullets. The antihydrogen ions would be accelerated by a lightweight, low-power, cheap 200 MeV neutral antiparticle beam weapon accelerator. They would be aimed at the target by the director lens, have their extra positrons stripped off in the stripping chamber, and mosey on their way to the target at a mere half of the speed of light. When they arrived at the target, however, they would hit with the energy of 2 GeV particle. They would annihilate with a proton in the target and release a spray of highly penetrating pions and gammas rays with a total energy of 2 GeV. The use of antimatter as bullets in a neutral particle beam weapon could easily cut the cost of making and deploying such weapons by a factor of 5 or 10. The savings would more than repay a research program to develop methods for making small amounts of antihydrogen, which are all that would be needed. We are not talking about grams of antimatter needed to make an antimatter bomb, or even the milligrams of antimatter needed for space propulsion. If a few thousand joules of radiation can kill a missile, then a few tens of nanograms (billionths of a gram) in each neutral particle beam weapon is all that's needed. No one has yet decided if neutral particle beam weapons are among the weapons that ought to be deployed by the SDI. If they are, however, a single microgram of antihydrogen would supply all the bullets needed for an entire constellation of neutral antiparticle beam weapons.

Space Planes

Antimatter is a manufactured fuel and is expensive compared to other fuels. It is unlikely that mirror matter will be a cost-effective fuel for power and propulsion except in space. Any fuel is expensive in space because it must be lifted out there first. However, if antimatter can be made for less than $10 million per milligram, mirror matter will be cost-effective for space propulsion and for power where solar energy cannot be effectively used. At the present subsidized price of a Space Shuttle launch, it costs about $5 million to put a ton of anything—including rocket fuel—into low Earth orbit. One milligram of mirror matter produces the same amount of energy as 20 tons of the most energetic chemical fuel available. So $10 million of mirror matter is actually a more cost-effective fuel in space than $100 million of chemical fuel.

The benefits don't stop there, especially when mirror matter is used as an energy source for a rocket. For missions around the earth and throughout the solar system, it's not necessary to use equal amounts of matter and antimatter to power a rocket vehicle. The best arrangement is to annihilate a few milligrams of antimatter with a few milligrams of normal matter, and use that energy to heat tons of reaction mass—water, methane, or hydrogen—which becomes a hot gas for the rocket to spew out the exhaust nozzle. Only one size of vehicle is then needed, no matter how difficult the space mission. Direct launch to geosynchronous Earth orbit, inspection of a satellite and return to orbital base, orbit of the Moon, or even a manned mission to Mars—all would be possible with the same ship! The vehicle will be more like an aerospace plane than like present rockets with their huge propellant tanks topped by a tiny payload. The antimatter-powered aerospace plane will carry about twice its dry weight in propellant and a small amount of mirror matter. If the mission is easy, it will carry only a few milligrams of mirror matter and use it to heat the propellant to a hot gas to carry out the mission. If the mission is difficult, it will take along a little more mirror matter. The ship will use it to heat the same amount of propellant to a very hot plasma (a kind of ionized gas, the stuff of which stars are made). The plasma exhaust will come out of the rocket with a much higher velocity. The mirror matter-powered aerospace plane can maneuver at will for days around the Earth, or go all the way to the Moon or Mars without having to add on extra fuel tanks.

A good way to explain this is with an example. There is an obvious future peacetime military space mission that would be very desirable to carry out, one that's "impossible" using present propulsion technology. The scenario is simple. We have a space station in polar orbit around the Earth. Most of the work on Byrd II Station is scientific in nature, carried out by civilian researchers from the United States, Canada, Europe (including Yugoslavia and Hungary, which have recently joined the Common Market), Australia, and Japan. The station also has a small contingent of U.S. military personnel who work in several free-flying modules not far from Byrd II. The station receives an urgent distress call from the Soviet Union's Sagdeev Station, which orbits the poles in the direction opposite to that of Byrd II. Two workers have been severely injured by a serious explosion while building new additions to their station. Sagdeev Station's shuttle has been knocked out of commission by the explosion. The workers need extensive medical attention, and it's just not available there. The Russian resupply shuttle could be made ready for launch in about 24 hours, but that might be too late.

Fortunately, the military crew at Byrd II have their own aerospace plane floating nearby. They are asked to leave Byrd II, rendezvous with Sagdeev Station, and take the injured workers on board. If their injuries are serious enough, the American space plane will return them to Earth. Otherwise, the space plane will bring them back to Byrd II for treatment at the fairly extensive medical facilities there. This is the kind of mission that the military frequently carries out during peacetime (and during wartime, too, for that matter).

This very simple-sounding double-reverse orbital mission cannot be done from space using chemical fuels. Low Earth orbital velocity is 7.7 kilometers per second. To reverse orbits within much less than half an orbital period or much less than 45 minutes requires a vehicle able to accelerate at greater than one G to reach a velocity of over 15 km/s. Chemical rockets can exceed one G, but they can't carry enough fuel. Aerobraking (by diving in and out of the atmosphere) might enable a chemical rocket aerospace plane to carry enough fuel, but it can take a lot of time, and time in this scenario may be critical. Nuclear and electric rockets can reach 15 km/s, but they can't accelerate anywhere near one G. Only antimatter rockets can do the job.

Then, to come back to the space station after the rescue requires that the rocket be able to reach a total velocity of about 31 km/s. To carry out this entire mission, a rocket using liquid oxygen/liquid hydrogen would require 500 tons of fuel for each ton of vehicle. That's an impossible requirement, since the tanks to hold 500 tons of fuel would themselves weigh more than a ton. However, this mission could easily be done by an aerospace plane carrying a few tons of propellant and a few tens of milligrams of mirror matter. Mirror matter makes this "impossible" peacetime military mission possible. This same kind of double-reverse orbital maneuver could also be useful in a not-so-peaceful situation: for dogfighting in an orbital space battle.

From Theory to Practice: A Plan

Mirror matter-powered space planes, particle beam weapons powered by antimatter—it sounds like a science fiction story. The Defense Department is not so sure. In fact, the U.S. Air Force is actively looking at mirror matter and how it might be used in the future.

It began in 1985. A consortium of scientists and engineers prepared a proposal to the SDIO (the Strategic Defense Initiative Office) for a program that could have led to a proof-of-concept demonstration of antiproton propulsion. The consortium included investigators from the Rand Corporation, Boeing Aircraft, United Technology, Hughes Aircraft, McDonnell-Douglas Astronautics, NASA/Jet Propulsion Laboratory, Los Alamos National Laboratory, the Air Force Astronautical Laboratory, and the Lawrence Livermore National Laboratory. Additional backers came from Rice University, Texas A & M, the California Institute of Technology, Fermilab, MIT, Rockwell International, University of Georgia, Kent State University, Case Western University, UCLA, and NASA/Ames Research Center. The consortium wanted to produce a demonstration of a subscale laboratory model rocket engine system based on antiproton annihilation. The idea was to do this by 1995, ten years after the start date. The proposed program had four major tasks: the production of antiprotons, handling and storage of antimatter, the interaction physics of the annihilation process, and the development of the propulsion technologies to handle the high-energy densities expected.

The first task, production, was aimed at developing a series of progressively larger antiproton sources within the United States. They would first produce the small quantities of stored antiprotons needed to support the fundamental and applied research work in the initial phases. That would be followed by larger quantities of stored antihydrogen for the propulsion engineering development work in the later phases.

The program would start with non-military basic research using the trillionth of a gram (picogram)-per-day facilities at CERN in Europe. The next phase would involve the development of facilities in the United States that could produce a nanogram per day. The machines to do this could be at either Fermilab or the Brookhaven National Lab, where antideuterium was first discovered. Next would be a facility to make a millionth of a gram (microgram) per day. This would involve taking an existing design for a high-current, low-energy proton accelerator and upgrading it to the higher proton energy needed to make antiprotons. A good candidate might be the Los Alamos National Laboratory HADRON machine (Figure 9–5). With sufficient funds, the production of mirror matter could go from picograms

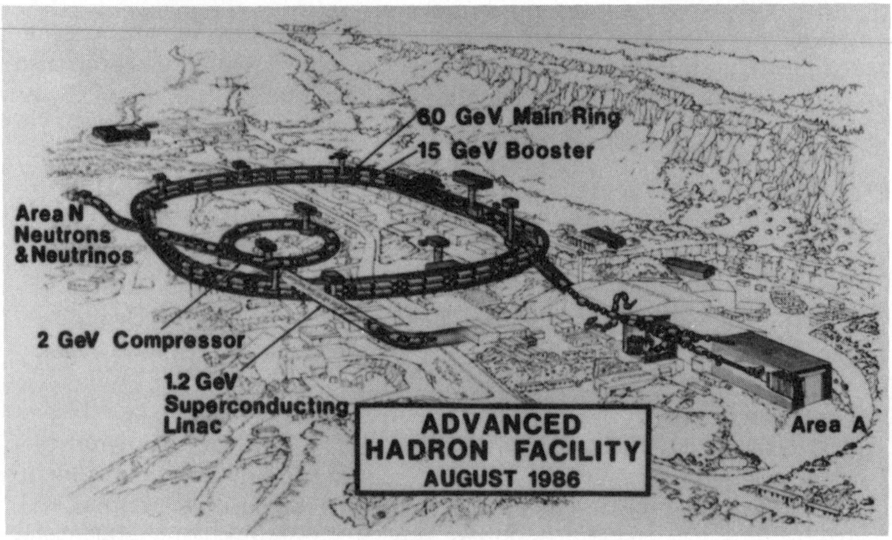

Figure 9-5 The Los Alamos National Laboratory HADRON machine is one candidate for an antiproton "factory" to make a microgram of mirror matter per day. *(Courtesy Los Alamos National Laboratory)*

to milligrams per day—a million-fold increase!—in several steps over ten years.

Antimatter handling and storage was the second task. It would be demonstrated early in the program that antiprotons can be captured and stored in a compact container, extracted from the container with minimal loss, and directed to any desired position required for their use in propulsion. The first part of this task, low-density storage of antiprotons in a compact container, has already been accomplished. Gerald Gabrielse and his coworkers have used a Penning trap to capture antiprotons coming from the Low Energy Antiproton Ring at CERN in Switzerland.

The proposed high-density storage effort would build upon the low-density work. It would develop techniques for slowing, trapping, storing, extracting, and feeding the quantities of antiprotons that would be needed for propulsion. The planned approach would be to form a compact mass of antihydrogen, which could then be suspended in either a magnetic or an electrical field trapping region. Much of this work could be simulated by using normal hydrogen, with antihydrogen only used to verify the process. The conversion of antiprotons to antihydrogen would be the key research and technology area. That would lead to the successful and credible demonstration of an antiproton propulsion engine.

The third task, interaction physics research, was separate from the propulsion development task because of its overall importance to the feasibility of antiproton propulsion. Also, many of the sub-tasks would involve fundamental physics research rather than engineering. This third task would involve a detailed study of the interaction of various forms of available antimatter with the matter and the magnetic fields likely to be used in an antiproton rocket. Most matter-antimatter physics happens at very high energies, because that's how physicists make new subatomic particles. Those reactions are well studied. The way antimatter reacts with normal matter at low energies is not so well known. At high energies these reactions are dominated by nuclear interactions. At low energies, though, the reactions are dominated by the electrical attraction of the antiprotons for the protons and the positrons for the electrons. There would be major differences in the speeds with which antihydrogen atoms and antihydrogen molecules would interact with normal molecular hydrogen, for example.

The final task, propulsion development, would tie together all the science and technology into a demonstration of a laboratory model of an antiproton rocket. The model would have all the characteristics of a rocket system suitable for various space operations. The first two years of work on this task would be a study of four sample missions directly related to SDI. They would include boost-phase interception of an enemy missile, midcourse-phase interception, Earth-to-orbit transportation, and trans-atmospheric flight for tactical warfare and strategic defense (that is, the aerospace plane). The next five years would involve engineering of the rocket subsystems, using knowledge gained from the interaction physics research task, and antiprotons from the production and storage tasks. Researchers would develop a "combustor" to investigate the interaction physics in a device with a volume resembling that of a typical rocket chamber. This would give an early look at exactly how and in what forms the energy is released from the annihilation of different forms of antimatter with different working fluids. Researchers would also need to develop a rocket nozzle capable of controlling and withstanding the high temperatures and pressures expected in an antimatter rocket. These could be many times higher than experienced by the Space Shuttle main engines. However, the exhaust from an antimatter rocket will be highly ionized, which might make the task easier, since magnetic fields can be used to partially protect the nozzle wall.

Another component that would have to be developed during this phase is the vehicle's fuel tank and propellant feed line. Researchers would need to develop a method of selecting a tiny antihydrogen ice crystal in the fuel tank, and extracting it from the tank. Then the crystal must be sent down tubing that maintains the temperature and pressure difference between the ultracold vacuum in the fuel tank and the roaring hell in the combustion chamber. Finally, researchers woud have to develop a method of injecting that tiny ice crystal into the combustion chamber so that most of the annihilation process takes place at the proper point in the chamber's interior.

No one said building an antimatter-powered rocket was going to be easy.

If the initial proof-of-concept demonstration were successful, a number of different possibilities for a factory capable of producing milligrams of antiprotons per day, or almost a gram per year, could be explored. Some of these designs are self-powering—they would not need an outside source of energy. Others could be built in space or on the Moon where there's no need to dig tunnels or install vacuum pipes. There is plenty of vacuum in space, and on the Moon. There's also plenty of energy in space. At the distance of the Earth from the Sun, the Sun delivers over a kilowatt of energy for each square meter. In the near term, solar electric power systems could be expected to be about 20 to 30 percent efficient. That means they would deliver about 200 to 300 watts of power per square meter of solar energy collector. A collector array 100 kilometers on a side would provide a power input of up to three terawatts (3,000,000,000,000 watts). That's enough to run an antimatter factory at full power, producing a gram of antimatter every week.

The proposal was ambitious. In one form or another, it will eventually happen. Unfortunately, its proponents could not see a way to speed things up. They could not bring the proof-of-concept demonstration—the firing of a laboratory-scale antimatter rocket with real antimatter—any closer than ten years from the start date. The SDI Office is a bureaucratic creature. It serves a President with a maximum term of eight years, and a Congress that often changes its mind. The SDIO couldn't wait until 1995. And by the time they got the proposal from the consortium, the Office didn't have the money to fund it. So the SDI will not include the consideration of antimatter in any of its phases, but will depend upon chemical space propulsion.

The SDIO's official position on antimatter was stated by J. David Martin, the SDI's Director of External Affairs, in a letter to one of the authors:

The Defensive Technologies Study of 1983 (the "Fletcher Report"), which formed the technical basis of the Strategic Defense Initiative (SDI), investigated antiproton beam weapons. The conclusion was that this area warranted further study. Later work by several laboratories showed that antimatter physics can have a number of high-payoff benefits for national security, but that antiproton beams did not appear to offer any advantages over neutral particle beams or to provide any specific strategic defense possibilities in the time frame of fundamental interest to the SDI (1995-2010).

The SDI has continued to fund some basic physics analysis in this area and will continue to consider proposals. However, I doubt it will be a high-priority funding item until such time as fundamental containment and production questions have been experimentally addressed. I believe the priority placed on this by the Air Force will assure that this work is adequately funded.

This would seem to be the final word on mirror matter and the SDI . . . at least for now. But the "mirror matter alliance" may change that.

The Mirror Matter Alliance

The scientific and engineering interest in mirror matter is broad and grow-ing. Particle physicists at CERN in Switzerland, the Institute for High Energy Physics (IHEP) in the Soviet Union, and Fermilab in the United States are all upgrading their antiproton production facilities to aid in their quest for Nobel Prizes. At the same time they're allowing some preliminary engineer-ing experiments critical to the understanding of the problems of storing antiprotons and the generation of mirror matter. A worldwide "antimatter alliance" is coming together. Their activities are creating the future:

- Gerald Gabrielse has already taken his cryogenic traps across the ocean to the LEAR machine at CERN in Switzerland to capture antiprotons. One day he may return to Harvard Yard carrying mirror matter through cus-toms. (The antiprotons should not have any problems getting into the country. They already have their passports. CERN's Antiproton Accu-mulator Ring, in which the antiprotons are collected, is built partly in Switzerland and partly in France. The antiprotons will have passed over international borders billions of times already!)
- Atomic scientists are planning to use antiprotons to study their interaction with normal matter atoms, sure that they will find uses for them in basic atomic research such as the internal chemical analysis of solid-state materials. Physicians are teaming up with particle physicists to study the use of low-energy antiproton beams for finding and destroying tumors deep inside a human body.
- Scientists and engineers at Los Alamos National Laboratory are busy with a number of interlinked studies to generate antiprotons with an upgraded particle accelerator, the HADRON. They will collect them in magnetic traps, study their interaction with normal matter atoms, and design magnetic chambers and nozzles to extract energy from the high-temperature plasmas produced by the interaction of mirror matter with normal matter.
- Giovanni Vulpetti at Telespazio, Rome, Italy, has founded a new Antimat-ter Research Team (ART) to carry out research on antimatter propulsion. The team is studying antiproton and positron production and storage, and antimatter engine simulation. Not too far in their future plans is some-thing like a technology demonstration.
- Systems analysts at the RAND Corporation are looking at the space system implications of the availability of mirror matter and its potential for drastically cutting the mass in orbit needed to carry out the goals of the Strategic Defense Initiative.
- The Air Force has been supporting studies at national laboratories and industrial research facilities on advanced propulsion using mirror matter. During late 1985 and early 1986 the Air Force carried out a special study called Project Forecast II, which defined what the Air Force should be

doing now in order to accomplish its projected missions in the twenty-first century. A team of 175 military and civilian experts evaluated more than 2,000 ideas for possible future technologies, many of them quite unconventional. In the end, the Forecast II study recommended that the Air Force work on 70 different areas, from "intelligent" weapons to integrated cockpit designs. Mirror matter made the final cut. The unclassified Executive Summary report states:

> We are … enthusiastic about an admittedly high-risk search for ways to use antiprotons. These unusual particles, currently produced at several locations throughout the world, will, when combined with protons, release enormous amounts of energy—far greater than that produced from any other energy source. If antiprotons could be stored and then combined with protons in a safe manner, we would have an incredibly rich energy source available for propulsion and power.

In a television interview on "Good Morning America" on February 18, 1986, General Lawrence A. Skantze, head of the Air Force Systems Command, gave a brief summary of the results of Forecast II. General Skantze specifically mentioned the trans-atmospheric vehicle (the "Orient Express") and the use of antimatter for propulsion. President Reagan has already publicly endorsed the "Orient Express." Is mirror matter next?

Perhaps. The ARIES office at the Air Force Astronautics Laboratory was a direct result of Project Forecast II. The office is now funding several antimatter studies that would lead to antimatter rockets for the military—and for NASA and anyone else as well. So far, all the research is unclassified. ARIES is also sponsoring a series of workshops on antiproton science and technology, which bring together scientists and engineers from all over the United States to work on developing a national scientific and engineering consensus on the need for a source of low-energy antiprotons in the United States.

• ARIES is not the only Air Force bastion in the Mirror Matter Alliance. The Air Force Office of Scientific Research (AFOSR) at Bolling Air Force Base outside Washington, D.C. has started a new program on antimatter research under the Physical and Geophysical Sciences Branch. The AFOSR is the primary source of basic research funds for the Air Force and is sponsoring university efforts on antimatter research, including the recently successful attempt by Gabrielse and his coworkers to trap antiprotons from the Low Energy Antiproton Ring at CERN in Europe.
• The Air Force Astronautics Laboratory, with financial support from the Air Force Office of Scientific Research, has also been supporting studies of advanced propulsion using mirror matter at the Lawrence Livermore National Laboratory and the Hughes Research Laboratory. Hughes has started an in-house program to study the slowing and cooling of molec-

ular hydrogen with laser light sources. This will complement a similar effort on atomic hydrogen at Hughes.

- The AFAL interest in propulsion is augmented by studies of mirror matter-powered missions to the outer planets by engineers and mission analysts at NASA's Jet Propulsion Laboratory. Similar mission studies and preliminary design studies of mirror matter-powered rocket engines are underway at Aerojet General, Boeing Aerospace, McDonnell-Douglas, and United Technology Research Center.

The Mirror Matter Alliance consists primarily of optimists. But they are *intelligent* optimists. They have better things to do than to waste professional time on something that will not work. They're convinced that making mirror matter is feasible. The real question is cost. Mirror matter at the present cost of $100 billion per milligram has already proven cost-effective for winning Nobel Prizes. Mirror matter at $1 billion per milligram might be cost-effective as an explosive "bullet" in a space-based mirror matter neutral antiparticle beam weapon. When the cost of mirror matter starts to drop below $10 million per milligram, then many new applications will come to the fore. At that price, mirror matter will be a cheaper source of energy in low Earth orbit than chemical fuel and possibly even cheaper than nuclear fuel.

Mirror matter is not a magical answer to every energy requirement in the panoply of Defense Department and NASA space missions. Even with adequate funding, it will take decades of hard work before mirror matter is available in quantity. Yet it is important that the planners and the thinkers at NASA and the Department of Defense, as well as the public at large, know that mirror matter is no longer imaginary matter invented by science fiction writers, but something real, being worked on by real people in a diverse and growing array of real organizations, that could have a significant impact on future military and civilian operations in space.

Chapter Ten

Opening Up the
Solar System

One of the most irreverent and creative performing groups of the 1960s and 1970s was the Firesign Theater. They did radio theatre (a nearly lost art), and their specialty was weird satire. With Lyndon Johnson, Richard Nixon, Vietnam, and Watergate—not to mention Woodstock, the drug counterculture, and sexual liberation—it was a time that was ripe for their brand of zaniness. Much of their humor can still be enjoyed on record albums, one of which is entitled "Everything You Know Is Wrong."

That could be the subtitle of this chapter. For once mirror matter becomes economically available as an energy source for space propulsion, then everything that space mission designers know about designing space missions will be wrong. Many currently implicit assumptions about planning robot or human-crewed missions in space will no longer be valid (see Figure 10–1). As we saw in Chapter 7, designers of space missions will have to develop new assumptions. Mirror matter changes the rules of the space travel game. Things that were once impossible will become easy.

Today's Plans for Tomorrow

Let's begin our exploration of mirror matter in space with a look at some of the plans NASA already has on the drawing boards for future space exploration. It's true that the Challenger disaster temporarily put the Shuttle program on hold and re-ignited the long-smoldering debate about the "usefulness" of having humans in space. Also, chronic budget cutbacks have severely curtailed NASA's once-ambitious plans to explore the solar system with robot space probes. But NASA does indeed have plans for the future, despite past tragedies and current soul-searching.

One mission due to be launched by 1989 is the Galileo space probe to Jupiter. After several years of orbiting the Sun and swinging past Earth to

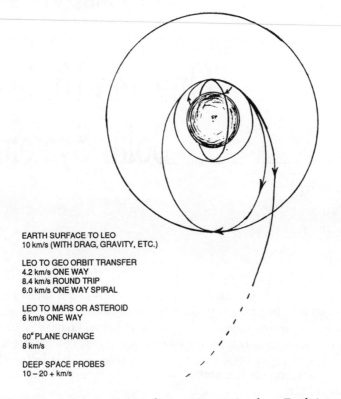

EARTH SURFACE TO LEO
10 km/s (WITH DRAG, GRAVITY, ETC.)

LEO TO GEO ORBIT TRANSFER
4.2 km/s ONE WAY
8.4 km/s ROUND TRIP
6.0 km/s ONE WAY SPIRAL

LEO TO MARS OR ASTEROID
6 km/s ONE WAY

60° PLANE CHANGE
8 km/s

DEEP SPACE PROBES
10 – 20 + km/s

Figure 10-1   Space missions first require getting from Earth into orbit. Interestingly enough, it is easier to go from low Earth orbit (LEO) to almost anywhere else than it is to first get from Earth into low Earth orbit. Once you have gotten into orbit, in other words, you are already halfway to anywhere. (*Courtesy Maj. Gerald Nordley, USAF*)

get a gravity-boost to its velocity, Galileo will finally orbit Jupiter in the late 1990s. It will also drop a probe into the jovian atmosphere. Another mission scheduled for launch as soon as the Space Shuttle is operating again is the Magellan space probe to Venus (Figure 10–2). This is the "Venus Radar Mapper" project. Magellan will be partly built from spare parts from Voyager and other probes. It will orbit Venus and use a powerful radar system to map the surface of the planet with an accuracy never before possible. Both Galileo and Magellan were planned many years ago, but were delayed by budget cuts and then by the Challenger disaster.

During the early 1980s NASA's Solar System Exploration Committee drafted a set of modest goals for exploration of our solar system. These relatively inexpensive programs would use two kinds of space probes to keep costs down and still get good scientific data back. The first would be commercially available Earth-orbiting spacecraft modified for use as "Planetary Observers," a very inexpensive way to do some missions in the

Figure 10-2   The Magellan space probe will be launched in 1988 or 1989 on a mission to map the surface of Venus with high-powered radar beams. *(NASA illustration)*

inner solar system, to Mercury, Venus, the Moon, Mars, and some asteroids that come near the Earth.

The second type of space probe would be a new kind of modular spacecraft called the Mariner Mark II. It would have a basic "bus" or central core that would be customized with add-on parts for specific missions. Mariner Mark II spacecraft would be ideal for missions to the asteroids and comets and for return missions to the outer planets of Jupiter, Saturn, Uranus, and Neptune (Figure 10-3).

One Planetary Observer mission has already been approved by Congress. The Mars Observer space probe will be launched in 1992 (possibly by an expendable booster rather than the Space Shuttle), and will begin exploring Mars in early 1993. The Mars Observer probe will orbit the Red Planet for at least one martian year (about two Earth years). It will gather information about the martian weather and climate, and also use remote-sensing devices to look for the presence of water beneath the martian surface. The Mars Observer will be a small spacecraft, weighing only a few hundred kilograms at the most, with only a few scientific instruments.

No Mariner Mark II missions are yet off the drawing board. The one most

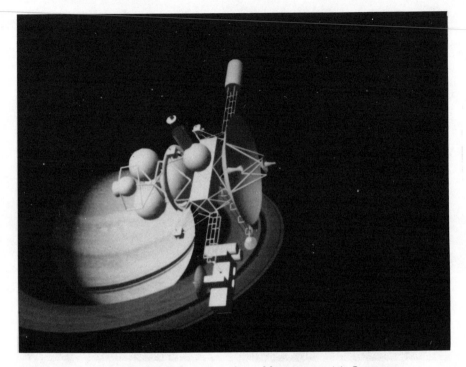

Figure 10-3   A Mariner Mark II spacecraft could return to visit Saturn on the Cassini mission. Mariner Mark II probes could also be used for a Saturn Ring Rendezvous/Sample Return mission, but the propulsion requirements are very high. Mirror matter propulsion would make such a mission easier. *(NASA illustration)*

likely to be approved first will be a mission to a comet and an asteroid. The Comet Rendezvous/Asteroid Flyby (CRAF) mission would use a Mariner Mark II-type spacecraft to fly past two asteroids and then rendezvous with a comet while it approaches the Sun. If a CRAF were launched in 1992 (Figure 10-4), it would first spend two years flying around the Sun. In June 1993 it would fly past the asteroid Malautra, taking pictures and examining it with other remote sensors. The space probe would then zip past the Earth in 1994 and get a gravity-boost to increase its velocity. In January 1995 CRAF would fly past the asteroid Hestia and examine it with cameras and other instruments during the few moments of closest approach. Finally, in October 1996 CRAF would rendezvous with Comet Tempel 2. It would spend nearly three years flying in formation with the comet, making lengthy studies. CRAF might even drop a small penetrator into the comet itself, to study the chemical and physical properties of the nucleus—the first landing of a human-built probe of any kind on a comet.

CRAF and other Mariner Mark II space probes would be larger than the Planetary Observers. They would carry about 130 kilograms of instruments, as well as nearly 3,000 kilograms of propulsion fuel to make changes in

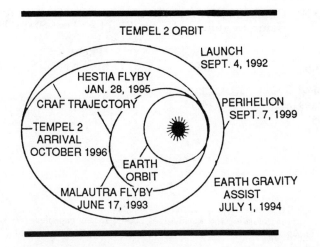

TEMPEL 2 ORBIT

LAUNCH
SEPT. 4, 1992

HESTIA FLYBY
JAN. 28, 1995

CRAF TRAJECTORY

PERIHELION
SEPT. 7, 1999

TEMPEL 2
ARRIVAL
OCTOBER 1996

EARTH
ORBIT

MALAUTRA FLYBY
JUNE 17, 1993

EARTH GRAVITY
ASSIST
JULY 1, 1994

Figure 10-4 One near-future space mission is CRAF, Comet Rendezvous and Astroid Flyby. This drawing shows one version of the mission that could be launched in 1992. From Draper, in *Aerospace America*, October 1986, p. 30. *(Copyright © American Institute of Aeronautics and Astronautics; reprinted with permission)*

path and speed. The mission velocity for CRAF (that is, its delta-v or total changes in its velocity) will be about 2.5 kilometers per second. In fact, that's the general mission velocity necessary for orbital missions to Mars, as well as flyby or rendezvous missions to comets and asteroids.

Obviously, such missions can be carried out with current space propulsion technology. But these missions are also limited ones. Because they are dependent on storable chemical propellants, the spacecraft have deliberately been designed to be relatively small in mass, and so they cannot carry a lot of scientific instruments. Orbital missions to the giant planets, like Jupiter and Saturn, will call for probes that carry large amounts of fuel so they can slow down and go into orbit. That means even fewer scientific instruments can be carried. The Galileo space probe to Jupiter is about the limit of what can be done with current space propulsion technology. The spacecraft is not much bigger than the two Voyagers that preceded it to Jupiter.

Still, space scientists continue to explore possible missions to the planets that could be carried out before the twenty-first century (assuming NASA gets the money from Congress). They include sample return missions to a comet, rendezvous missions with near-Earth and Main Belt asteroids, new missions to the surface of Mars, Saturn orbiters, and Titan flybys and probes (Tables 10–1 and 10–2). The space probes for most if not all of these missions would have on-board propulsion systems that use solid fuel propellant with

Table 10-1   Comet Flyby/Sample Return mission opportunities
(Launch vehicle: Space Shuttle/Two-Stage IUS)

Target (apparition)	Launch date	Flight time, yr	Flyby V_∞, km/s
Borrelly (1994)	3/91	4	18
Tempel 2 (1994)	4/92	3	11
Tuttle (1994)	10/92	3	33
TGK (1995)	1/94	4	9
HMP (1995)	3/94	2	12
Wild 2 (1997)	11/96	3	10
Tempel 2 (1999)	3/99	3	11

Notes: One-day launch period.
One-stage Earth-storable propulsion ($I_{sp} = 310$ s).
Spacecraft net mass of 100 kg includes 200-kg return capsules, 150 kg for sample collector/dust shield, and 50 kg for interface requirements.

Source: Wallace, et al., in *Journal of Spacecraft and Rockets*, May 1985, p. 318. Copyright © *American Institute of Aeronautics and Astronautics; reprinted with permission.*

Table 10-2   Near-Earth asteroid rendezvous mission opportunities
(Launch vehicle: Space Shuttle/Two-Stage IUS)

Target	Launch date	Flight time, yr	Total ΔV, km/s
1980AA	1/91	1.95	1.68
Anteros	5/92	1.40	1.13
Ivar	8/93	2.10	1.50
1980PA	11/94	1.80	1.86
Anteros	5/95	2.00	1.58
Eros	1/96	1.91	1.89

Notes: Data for one-day launch period.
One-stage Earth-storable propulsion ($I_{sp} = 310$ s).
Total ΔV includes 0.115 km/s for interplanetary navigation and post-rendezvous maneuvers.

Source: Wallace, et al., in *Journal of Spacecraft and Rockets*, May 1985, p. 319. Copyright © *American Institute of Aeronautics and Astronautics; reprinted with permission.*

an exhaust velocity of about 3 km/s, for changes in velocity after a probe is on its way to the target. One study done at the Jet Propulsion Laboratory assumed that the space probes would have a mass of no more than 600 kg, excluding the mass of the propulsion system and fuel. A Titan Flyby or Probe mission would carry an additional 200-kg probe similar to the one for the Galileo mission, so it would mass about 800 kg. A mission to gather a sample of a comet and return it to Earth would have an additional 400 kg of payload. That would include the sample-collection device and the Earth-return capsule. The comet sample return mission spacecraft would therefore mass about 1,000 kg.

The Challenger disaster pretty much wiped out all the optimistic time-lines for future robot exploration missions through the end of the century. The first new mission to a planet, the Mars Observer mission, will not be launched until 1992. CRAF, even if it is approved by now, will not be launched until 1994 at the earliest. It is still possible, though, that a few of the scenarios planned for the last few years of the twentieth century could become reality:

- A relatively simple but effective martian mission is the Mars Surface Probe mission. A modified Planetary Observer spacecraft would carry several penetrators, meter-long probes which look like golf tees. The main spacecraft would drop them as it approached Mars. They would enter the martian atmosphere and fall to the surface of the planet. The penetrators would be targeted for interesting locations—the martian ice caps, for example, or one of the giant volcanos like Olympus Mons, or an area that was once scoured by flowing water. Mars Penetrator missions could be launched just about every other year after 1992 (that is, 1994, 1996, 1998) if the money were there and the probes were built. Regrettably, it isn't, and they aren't. Yet.
- It might still be possible to send Mariner Mark II space probes to several comets before the end of the century. Today's boosters could be used to launch rendezvous missions to Comet Encke in 1994, and to Tempel 2 in 1997. Sample-return missions could be sent to Comet Wild 2 in 1997 and to Tempel 2 in 1999.
- Missions to near-Earth asteroids before the turn of the century are also possible (near-Earth asteroids are tiny chunks of rock whose orbits lie very close to Earth's). Two possible rendezvous missions to near-Earth asteroids include launches in 1995 to Anteros, and in 1996 to the famous asteroid Eros.
- Most asteroids are in the so-called Main Belt, between the orbits of Mars and Jupiter. It might be possible to launch space probes on rendezvous missions in 1994 to the Main Belt asteroids Gisela or Euterpe. The mission velocity for a Euterpe rendezvous is 2.5 km/s; for Gisela it's 3.6 km/s. Both these missions would need to get a gravity boost from a close Mars flyby.
- Missions to Saturn that would orbit the ringed planet and perhaps drop a probe into the atmosphere of the giant moon Titan are barely possible

before the turn of the century. They'd all need some complex navigation, with several gravity boosts from Earth, Jupiter, or both. Such missions could be launched in 1995, 1996, 1997, or 1998. They'd all take four to five years for the spacecraft to reach Saturn, so in a sense these are really missions for the twenty-first century.

All of these robot exploratory missions assume the use of chemical propulsion systems for the space probes. If we look at more advanced propulsion systems, though, things get rather interesting. The best advanced propulsion system likely to be available in the next 10 to 15 years is probably *nuclear electric propulsion*, or NEP. This is essentially an ion propulsion system that uses a 150-kilowatt nuclear reactor as the source of its power. NEP is slow—but very powerful. The best chemical propulsion systems have exhaust velocities of about 5 km/s. NEP could achieve exhaust velocities of 40 to 70 km/s! The ion engines could run for as long as 20,000 hours, or over two years, if necessary.

NEP makes possible return missions to Saturn and Uranus, with space-craft that would actually orbit those planets instead of just flying past them like Voyager 1 and 2. Another mission NEP would make feasible would be an orbital mission to Neptune (which Voyager 2 will fly by in 1989) that would also drop a probe into the planet's atmosphere. NEP propulsion would even make it possible to send a space probe to Pluto, with a gravity boost from Jupiter. All but the Neptune mission would take less than six years from launch to arrival.

The space probes for these NEP-driven missions would be larger than the Voyager-class spacecraft, massing 950 kilograms or so (excluding the mass of the NEP propulsion system and its propellant). None of these kinds of missions are possible before the twenty-first century; NEP systems have not progressed beyond the drawing board and the computer screen, and will almost certainly not be a reality before the year 2001.

That's about as far as NASA and other space scientists and engineers have ventured into the future of robot space exploration. The future of human space travel beyond Earth orbit is even less certain at this point. The two most definite proposals for future human exploration of the solar system center on a return to the Moon, and on a human-crewed mission to Mars. Many space enthusiasts, including some scientists, have long urged that the United States return to the Moon and establish a permanent human base there. The Moon offers many attractive features to optical and radio astronomers, for example, for future science. The lunar surface and subsurface may also contain significant amounts of metals and other materials that could be commercially mined in the twenty-first century. Some visionaries even propose building giant greenhouses on (or beneath) the lunar surface, and turning our natural satellite into a giant future garden.

Until recently, NASA and the larger space science community didn't seem too excited about returning to the Moon. This could be a case of the "ho hums" and the "money blues." We've done it before, the first argument goes, so why do it again? The second argument is probably more to the point: It is possible the general tax-paying public would see a return to the Moon as a greater "waste of money" than the first visit. However, a special panel chaired by former Shuttle mission specialist Sally Ride has recommended that the United States return to the Moon and build a permanent lunar base. The proposal is now starting to gain more momentum, with even high NASA officials speaking out in its favor. It is entirely possible that Americans could return to the Moon and begin building a permanent base there by the end of this century.

Other people favor a human-crewed mission to Mars instead. Mars is fascinating; Mars is mysterious; Mars is—well, Mars! It is the famous Red Planet.

What's more, the Soviet Union makes no bones about seeing the Red Planet as "red" in more ways than color. They plan on going there. They would like to do it with someone else (namely, the United States), at least to cut their costs. Some argue that the Soviets also want to steal our space technology again, à la the Apollo-Soyuz Project of the late 1970s. However, the Soviets no longer have a big incentive to steal space travel technology from the Americans. The Soviet Union, like it or not, is now the leader in space activities.

For these reasons and others, many inside and outside the official U.S. space program advocate a human-crewed mission to Mars sometime in the early years of the twenty-first century. It could very well happen. Such a mission has been studied to exhaustion for decades. It will likely use improved versions of current chemical space propulsion technology. Soviet Energia-class heavy-lift boosters will probably put the hardware into Earth orbit. The spacecraft will be assembled at the current version of a Soviet space station. The United States, Japan, and probably France and Germany will furnish much of the advanced computers and artificial intelligence programming for the mission. The crew will be mixed by gender, race, and nationality. At least one or two Americans will get to go to Mars. So will several Russians, and no doubt a smattering of representatives from Third World countries.

The spacecraft will take the long route to Mars and back: nine months of coasting in a so-called Hohmann transfer orbit from Earth to Mars, a technique that saves propulsion fuel, and thus cuts down on the cost of the mission. Once there, the ships will go into orbit around the planet, and at least one lander with Russians and an American will descend to the sands of Mars. The crew will stay at Mars (on the surface and in orbit) for a few weeks or months; then they will leave orbit and return to Earth via a year-long journey—total trip time: about two years. The crew will return to the adulation of the entire planet, and a comfortable future living off the royalties of books and commercial endorsements.

No one is seriously planning any human-crewed missions to anywhere other than the Moon and Mars. There are no plans, casual or otherwise, for human exploration of the upper atmosphere of Venus in balloon-carried gondolas; no plans for human visits to the north pole of Mercury; no plans for human landings on (much less commercial mining of) the nickel-iron asteroid Eros, which sometimes comes within 17 million kilometers of Earth. There are no plans for the human colonization of Mars. Certainly there are no plans for sending human beings some six billion kilometers to land on Pluto. The exploration and colonization of the solar system by human beings is still science fiction, as far as NASA and the space science community is concerned.

Mass Ratio and a Mission

Now that we've examined some of NASA's current plans (and dreams) for the future, let's look at a human-crewed mission that NASA isn't thinking about because it's "impossible," says NASA. The space agency is right, but not forever.

Suppose you're an engineer designing a passenger-carrying space plane that will ferry people to and from different space stations and space cities in Earth orbit. Your passenger ferry will have to perform in basically the same way as the X-wing fighters in the movie *Star Wars*. It will need to be able to carry out double-reversals in orbit. In Chapter 9 we saw what a military space force could do with a space vehicle capable of such maneuvers, even in peacetime. Now let's look in closer detail at how to make it work in the first place.

Reversing direction in orbit and then re-reversing direction is not as easy to do as George Lucas portrayed. In fact, the simple maneuver of reversing orbital direction is nearly impossible with chemical rockets. It calls for canceling the spacecraft's initial orbital velocity, and then building it up again in the opposite direction. Low Earth orbital velocity is 7.7 km/s, with the spacecraft already in orbit. First it must cancel that velocity, and then accelerate to 7.7 km/s in the opposite direction. The total mission velocity is therefore 15.4 km/s—twice what was needed to get into orbit in the first place, and 4 km/s more than escape velocity.

Now, suppose the spacecraft needs to reverse direction a *second* time, and carry passengers back to its space station base. That means another 15.4 km/s of velocity change. So a spacecraft that can perform a double-reverse orbital maneuver has to be capable of at least 31 km/s of velocity—almost three times Earth escape velocity!

Nevertheless, you want to at least try to design a space ferry that can do these kinds of things. You know the mission velocity you need to achieve—15.4 km/s for a reverse orbit, and 31 km/s for a double-reverse orbit. You go looking for a propulsion system that can do it. First, you look at chemical propulsion systems using storable propellants—the kinds that can sit in the spacecraft's fuel tanks for long periods of time without any elaborate storage procedures. The best exhaust velocities from such propulsion systems are about 3 km/s. The mass ratio for a spacecraft capable only of a simple reverse orbit will be 175. That's much too high—175 tons of rocket fuel for every ton of spacecraft, pilot, and weapons. Can you do better?

Yes, a high-energy chemical propulsion system using liquid hydrogen and liquid oxygen has an exhaust velocity of 5 km/s. That changes your mass ratio (for a simple reverse orbit, anyway) to 22. If you're desperate, you might find that mass ratio worth trying. However, your space ferry will still be a huge beast, since a mass ratio of 22 is essentially equivalent to 22 tons of fuel and spacecraft for every ton of spacecraft and payload (that's pilot, passengers, and baggage). Is there any way to get the mass ratio even lower?

There is, but it means using a space propulsion system that has not yet been used: nuclear thermal propulsion (Figure 10–5). Such a system will

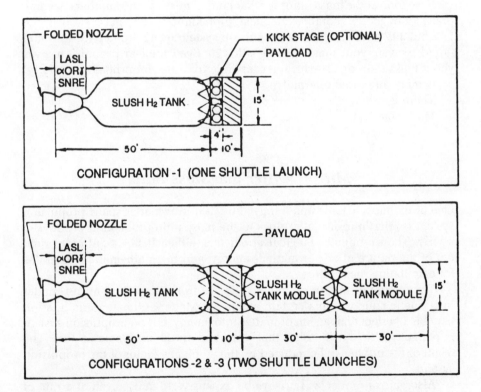

Figure 10-5 Nuclear rocket propulsion (NRP) is one method of solar system travel being studied by space scientists. This drawing shows two possible arrangements for NRP rockets. At the left in each drawing is the rocket nozzle and small nuclear reactor engine (SNRE). The slush hydrogen tank holds the liquid hydrogen used as working fluid for the rocket. The nuclear reactor would heat the liquid hydrogen to a high-velocity gas for thrust. In the top version, the tank is topped by an optional kick stage and then the payload. Larger versions of an NRP would have additional slush propellant tanks. One shuttle launch could carry the top configuration into orbit. At least two shuttle launches would be needed for the lower multi-tank configuration. *(From Niehoff and Friedlander, in* Journal of Spacecraft & Rockets August 1974, p. 569. *Copyright © American Institute of Aeronautics and Astronautics; reprinted with permission)*

have exhaust velocities in the range of 9 km/s. That means a state-of-the-art mass ratio of 6 for a space ferry that can do a simple reverse orbit maneuver. Some recent designs for particle-bed nuclear reactor rocket engines weigh close to what their chemical counterparts would weigh, including crew, passengers, baggage—and shielding. Other problems remain, though, including concern about the radioactive exhaust and passenger fears of the nuclear reactor.

The truth is, however, that a single-reverse orbit capability is really not going to be sufficient if your space ferry is to be truly effective. The space ferry will have to at least be capable of double-reverse maneuvers, and that means 31 km/s of velocity change at a minimum.

If we look at the mass ratios for that kind of mission, the numbers become pretty outrageous. Storable chemical propellants? Your mass ratio is not 175, but 30,700! If you go to a propulsion system using liquid hydrogen and liquid oxygen, your mass ratio is still 490. Even nuclear propulsion gives you a mass ratio of 32—still unacceptable for a decent orbital space ferry.

Is there any other alternative?

There is.

Mirror matter.

Mirror Matter and Mass Ratio

One of the most astonishing things about a mirror matter space propulsion system is what mirror matter does to the mass ratio problem: It destroys it, utterly. There's nothing magical about this, either. It flows naturally right out of the nature of mirror matter's energy, and the mathematical formulas for calculating mass ratios.

The energy of the rocket exhaust comes from converting mirror matter into kinetic energy with a certain efficiency. Matter-antimatter annihilation may be a perfect conversion of matter into energy, but no propulsion system is perfectly efficient in using that energy; there is always some loss. The result of $E=mc^2$ has to be reduced by the efficiency factor of the propulsion system.

Also, mirror matter will always be expensive to make, even at a rate of grams per year. Mirror matter must be economically competitive with other space propulsion systems, or it will never get off the ground. So we want to strike a balance between mirror matter costs and propellant costs.

With this in mind, some fairly simple mathematics show something very interesting. The amount of mirror matter for a space propulsion system is minimized when the rocket's exhaust velocity is 63 percent of its *mission characteristic velocity* (the sum of all the velocity changes needed for the mission). And this changes the way space mission designers will think about mass ratio. If we can optimize a spacecraft's exhaust velocity to 63 percent of the mission velocity, mass ratio becomes a constant for a mirror matter propulsion system. We plug it into the mass ratio formula found back in Table 7–3:

$$R = e^{V/v}$$

This means that R (mass ratio) equals the number e (2.718) raised to the power of 1 divided by 0.63 (the mission velocity V divided by the exhaust velocity v), or 1.59. Many pocket calculators today can do powers of numbers: It's then a matter of seconds to discover that $e^{1.59}$ equals 4.9. This leads to an inescapable, and astonishing, mathematical revelation: If we use the minimum amount of mirror matter necessary for our propulsion system, in order to keep costs as low as possible, the mass ratio for a mirror matter propulsion system *will never be worse than* 4.9. Ever. It doesn't matter how efficient we are at $E=mc^2$. The mission velocity is irrelevant. The type of reaction fluid we use is unimportant—hydrogen, water, or methane, it doesn't matter. This constant mass ratio for minimum mirror matter use holds for all conceivable missions in the solar system. It only starts to change for interstellar missions, where the mission velocity begins to approach the speed of light.

Now let's go a step further. Instead of minimizing just the cost of mirror matter, suppose we try to minimize the *total* fuel cost for our propulsion system. Remember, in a mirror matter rocket, the source of propulsion energy is separate from the propellant or reaction fluid. This isn't the case with chemical rockets, for example, where the source of the propulsion energy is the propellant, and the propellant is the reaction fluid. (The reaction fluid is what gets thrown out of the rocket in one direction [action] to cause the rocket to head in the opposite direction [reaction].) In a mirror matter rocket a few milligrams of mirror matter reacts with a similar amount of matter to create a huge amount of energy. That energy heats a large amount of reaction fluid. So the total initial mass of the rocket consists of the mass of the vehicle, the mass of the reaction fluid, and the mass of the energy source. Only a small part of that third component is the mass of the mirror matter. Antihydrogen, to be exact, will be the source of energy for propulsion. Something else—hydrogen, water, liquid methane, whatever—will be the working fluid or reaction mass thrown out the back of the rocket. What we want to do is reduce to a minimum the costs for reaction mass as well as mirror matter. This will result in a new optimum mass ratio. Water may be cheap when it's sitting on the ground, but it costs a lot of money to get it—or anything else—into orbit.

A cost analysis of this situation, though, brings us to some astonishing conclusions. Depending on the relative costs of mirror matter per milligram and reaction mass per ton, the optimum mass ratio for total minimum fuel cost is *still between 2 and 4*. That in itself is utterly astounding—space propulsion systems with mass ratios of 2! Still more amazing is the bottom line. Even under the most expensive conditions, the mass ratio never exceeds 5.

This compares very favorably with a liquid hydrogen/oxygen propulsion system, which can have a mass ratio of 5—but with the total mission velocity of less than 10 km/s. Its mass ratio will be greater than 100 at a mission velocity of 23 km/s, and will exceed 1,000 at 35 km/s. Mirror matter propulsion systems can reach mission velocities of more than 45 km/s and still have mass ratios of 5! (See Tables 10–3 and 10–4.)

Now: how does all this apply to our space ferry? Even with nuclear

Table 10–3   Mirror Matter and Mass Ratio

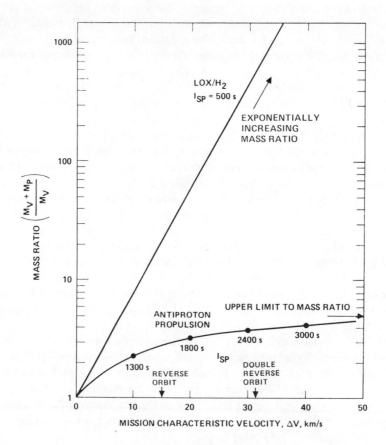

Mirror matter space propulsion destroys the limits of mass ratio. This graph shows that antiproton propulsion is immensely more powerful than even liquid hydrogen/liquid oxygen chemical propellants. For example, a "double-reverse orbit" mission with a Star Wars-type X-wing space fighter would require a mass ratio of more than 400 for even the best chemical propulsion system. A mirror matter-powered space fighter can perform such a mission with a mass ratio of about 4.

Table 10-4 Impossible Missions Made Possible

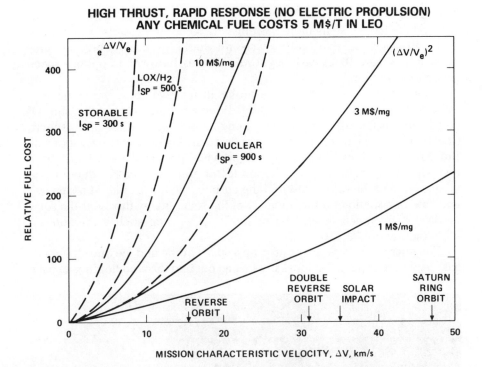

HIGH THRUST, RAPID RESPONSE (NO ELECTRIC PROPULSION)
ANY CHEMICAL FUEL COSTS 5 M$/T IN LEO

This graph shows how antiproton propulsion makes impossible missions possible. Any chemical fuel costs at least $5 million in Earth orbit. If mirror matter can be made and used for less than $10 million per milligram, it will become economically competitive with storable and cryogenic chemical fuels. If it costs less than $3 million per milligram it is even more economical than nuclear propulsion.

propulsion, it can only achieve a mass ratio of 6, and that's just for a ferry that can do a single reversal of its orbital direction. The best mass ratio for a space ferry with nuclear propulsion that can carry out a double-reverse maneuver is 32. However, the mass ratio for a spacecraft using a mirror matter propulsion system will never be worse than 5. What's more, a mirror matter propulsion system can have both high thrust and high specific impulse. To increase thrust or specific impulse, or both, we simply use a few more milligrams of mirror matter, thus increasing the energy of the propellant exhaust. Our ferry will be able to carry out orbital reversals with ease.

More "Missions Impossible"

The space ferry is only one "mission impossible" made possible by mirror matter. Let's look at some more, starting with some "impossible" robot space probe missions. The Star Probe spacecraft would fly within a million miles of the Sun (Figure 10–6); we'll begin with the possibility of dropping a space probe *into* the Sun.

This is not a waste of a space probe. In fact, we could get some very important scientific data by literally dropping a probe into the sun. Of course, the spacecraft won't survive. Long before it gets to the *photosphere* (the visible "surface" of the sun as seen from Earth) it will be vaporized. But during that long, final plunge to oblivion, protected for a while by a heat shield, the Solar Impact Mission (SIM) probe will gather data on the composition of the solar wind and the *corona* (the Sun's outer atmosphere), accurate information on the intensity of the solar energy flux near the solar "surface," and much more. The scientific gains from such a mission are more than worth its cost.

The question is, how does one drop a space probe into the Sun? It's not as easy as it may appear. One does not just go out into the void holding a space

Figure 10-6 The Star Probe spacecraft is a mission envisioned for the twenty-first century: a probe that will fly within a million miles of the Sun. *(NASA illustration)*

probe, and then let go of it. The orbital velocity of Earth must be canceled so the space probe can drop into the Sun. The reason? The Earth orbits the Sun at an average velocity of 29.6 km/s. Everything in orbit around Earth is still traveling around the Sun at the same velocity as the Earth. Even a spacecraft that has barely escaped from the Earth's gravitational grip by achieving a velocity of 11.2 km/s is still in orbit around the Sun, and essentially in the Earth's orbit at Earth's solar orbital velocity. Taking into account other forces affecting the SIM space probe, the mission characteristic velocity for this impossible mission is about 35 km/s.

One way to reach this velocity is to swing the spacecraft past Jupiter, and use that giant planet's gravitational force to cancel out the probe's velocity and send it dropping back into the inner solar system and finally into the Sun. However, this means going almost 750 million kilometers and many years in the wrong direction. At this point, that is the only way to drop a space probe into the Sun. Using a propulsion system with storable chemical propellants results in a mission with a mass ratio of 117,000. Using a liquid oxygen and hydrogen propulsion system gives a mass ratio of 1,100. And even nuclear propulsion means a mass ratio of 49—still too high.

So we turn to mirror matter propulsion. A mission velocity of 35 km/s is easy to achieve with such a propulsion system. Its specific impulse will be somewhere around 2,700 or 2,800 seconds (or an exhaust velocity of 27 to 28 km/s). Just as importantly, the space probe's total mass ratio will be about 4. There is no need to make a 750 million kilometer detour to Jupiter. SIM can start its long drop into the Sun from right here in Earth orbit.

Another "impossible mission" made easy with mirror matter is the Saturn Ring Rendezvous, or SRR. The spacecraft will not merely fly past the planet, like Pioneer 11 and Voyagers 1 and 2, but will actually go into orbit around Saturn. More than that, it will move into the rings of Saturn and take up an orbit within them. There it will study the ring particles close up, discovering exactly what they are made of, how big they are, and what their chemical properties may be.

This is another mission that has already been studied by scientists at the Jet Propulsion Laboratory. The mission velocity for SRR turns out to be 48 km/s. Once you, the mission designer, have defined the mission (SRR) and know its mission velocity (48 km/s), you go looking for a fuel which will have an exhaust velocity or specific impulse great enough to do the job. Storable chemical propellants? No way. With an exhaust velocity of 3 km/s, the mass ratio for this kind of mission would be—8,900,000!

What about liquid hydrogen and liquid oxygen, with—as we've seen—exhaust velocities of up to 5 km/s? Unhappily, that's still not enough. The mass ratio is now only 15,000, but one kilogram of deliverable payload per 15 tons of fuel is not acceptable.

Now you turn to really exotic propulsion system proposals. Nuclear propulsion? Sadly, no, not quite. Even with an exhaust velocity of 9 km/s, your spacecraft's mass ratio is going to be 200, and any spacecraft designer will assure you that a mission with even a 200 mass ratio is impossible to achieve. Of course, you could turn to ion propulsion. Thrusts are very low, but specific impulses and engine firing times are very high. You'd be able to achieve the needed total mission velocity of 48 km/s. JPL proposed using nuclear electric propulsion for this mission. It would work just fine. The kicker for such a mission, though, is time. It would take even such a highly advanced ion system more than eight years to get the probe from Earth to its final position in Saturn's rings. The JPL study hypothesized a SRR launch the day after Christmas 1992. The spacecraft would finally enter Saturn's rings on February 17, 2001. You'd like to get your SRR spacecraft to Saturn in a few years at most, and not in a decade.

With mirror matter, however, "everything you know is wrong." A spacecraft with mirror matter as the energy source will have a mass ratio of about 4.5 at a mission characteristic velocity of 48 km/s. The exhaust velocity for the propulsion system will probably be somewhere around 35 km/s—about seven times the exhaust velocity of the Space Shuttle's main engines—comparable to ion propulsion systems, but with thrusts hundreds of times as great. The SRR space probe would thus get to Saturn in a matter of months, not years.

Let's move still further out into the solar system. Voyager 2 made a flyby of the planet Uranus in January 1986, and will fly past Neptune in August 1989. Pluto, however, is not on Voyager 2's itinerary; its position in its orbit is such that not even the far-faring Voyager 2 can reach it. That doesn't mean we can't get there, however. In fact, a mission to Pluto could be combined with an even more ambitious project: an *interstellar precursor mission*. This is another deep-space mission that NASA's Jet Propulsion Laboratory has looked at, to be powered by the NEP system we talked about earlier.

One such Interstellar Precursor Mission (IPM) with NEP propulsion would also carry a Pluto Orbiter (Figure 10–7). The total spacecraft, IPM plus Pluto Orbiter plus propulsion, would have a mass of about 32,000 kilograms. With a second NEP booster included, that rises to 90,000 kilograms. That would include the propulsion system (Figure 10–8), nuclear reactors, and reaction mass. The actual IPM probe would mass only about 1,500 kilograms, including scientific instruments, computers, and other assorted electronics. The Pluto Orbiter itself would have a mass of 500 kilograms. It would also have a thousand kilograms of chemical propellant for its onboard rocket (Table 10–5). That gives the orbiter about 3.5 km/s of delta-v (or velocity change capability), more than enough to get itself into orbit around Pluto. The ultimate mass ratio for the IPM plus Pluto Orbiter is 45. For the IPM space probe itself the mass ratio is 60; for just the Pluto Orbiter it's 180. With current space mission planning assumptions, only the use of nuclear electric propulsion makes this kind of mission possible.

The IPM is a long-duration spacecraft, both in distance and in time (Figure 10–9). Its goal would be to reach a distance from the Sun of at least a thousand astronomical units (about 150 billion kilometers!) in 50 years. In fact, one recent version of the Interstellar Precursor Mission, devised by JPL scientists Aden and Marjorie Meinel, is dubbed TAU, for "Thousand

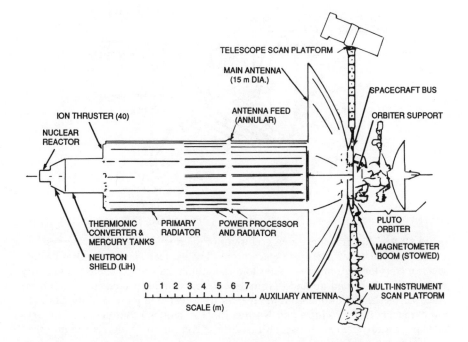

Figure 10-7 An interstellar precursor spacecraft, carrying a piggybacked Pluto probe, might look like this drawing. The spacecraft would use nuclear electric propulsion (NEP), with mercury for the working fluid and 40 ion thrusters all energized by a nuclear reactor. The craft's main antenna would be 15 meters (nearly 50 feet) wide, to communicate with Earth from nearly 150 billion kilometers (90 billion miles) away. *(From Jaffe and Norton, in* Astronautics & Aeronautics *January 1980, p. 43 Copyright © American Institute of Aeronautics and Astronautics; reprinted with permission)*

Astronomical Units." Using nuclear electric propulsion pretty much means a long travel time, since this is essentially an ion propulsion system powered by nuclear reactors. The booster version of the IPM would take four years to achieve a velocity of 25 km/s above solar escape velocity, and a distance of 1.2 billion kilometers from the Sun. The NEP booster would have been firing almost continuously during those four years. After it is jettisoned, the spacecraft's own NEP system would take over, burning nearly continuously for another eight years until the spacecraft reached a velocity of 150 km/s and a distance from the Sun of 155 AU, or 23 billion kilometers.

The Pluto Orbiter would be released from the main spacecraft very early in the mission. The Orbiter is somewhat like a boy on a skateboard hanging onto the bumper of a car that's slowly accelerating down Oak Grove Drive in Pasadena, California. The car is going from Pasadena to Seattle, Washington, but the boy just wants to go a few blocks from the main gate of JPL to the entrance of La Cañada High School. He lets go of the bumper very early in

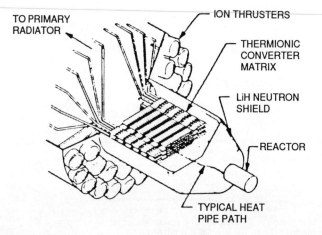

Figure 10-8   This drawing shows a closeup of the nuclear electric power system for an interstellar precursor space probe. The nuclear reactor supplies heat, which travels through the heat pipes to a thermionic converter. There the mercury working fluid is heated up and sent to the ion thrusters for propulsion. A neutron shield keeps stray neutrons from the reactor from reaching the rest of the craft and damaging it. *(From Jaffe and Norton, in* Astronautics & Aeronautics *January 1980, p. 41. Copyright © American Institute of Aeronautics and Astronautics; reprinted with permission)*

Table 10-5

Allocation	Mass, kg	
Spacecraft (without power and propulsion)	1200	
Pluto orbiter	1500	
Propulsion, dry (NEP, 500 kw$_e$)	8500	
Propellant (Hg): spiral from Earth orbit	2100	
heliocentric trajectory	18100	
Propellant tankage	600	
Total mass in Earth orbit, 1 stage	32000	
Booster stage (2 × 500 kw$_e$ NEP)		58000
Total mass in Earth orbit, 2 stages		90000

Performance	1 Stage	2 Stages
Distance at booster burnout, AU	—	8
V_∞ at booster burnout, km/s	—	25
Time from launch to booster burnout, yr	—	4
Distance at spacecraft NEP burnout, AU	65	155
V_∞ at spacecraft NEP burnout, km/s	105	150
Time from launch to spacecraft NEP burnout, yr	8	12
Spacecraft distance in 20 yr, AU	370	410
Spacecraft distance in 50 yr, AU	1030	1350

Source: Jaffe and Norton, in *Astronautics & Aeronautics*, January 1980, p. 41. Copyright © American Institute of Aeronautics and Astronautics; reprinted with permission.

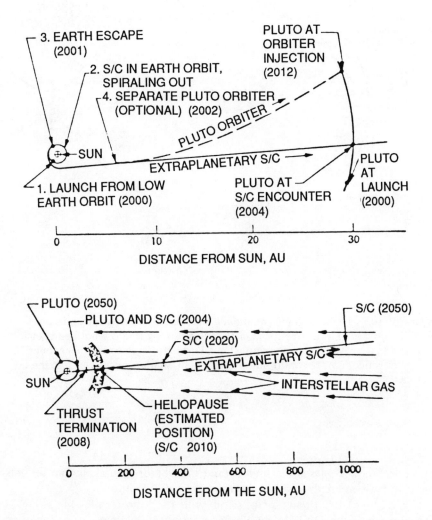

Figure 10-9 A suggested trajectory of an interstellar precursor mission. (Top) The spacecraft is launched in the year 2000, and spends a year spiraling out from Earth as the NEP engine accelerates it to escape velocity, which it reaches in 2001. A year later the Pluto orbiter separates from the main spacecraft and follows its own trajectory to Pluto. The main interstellar craft will pass Pluto in 2004, heading for interstellar space. The Pluto orbiter will reach the planet in 2012, and go into orbit. (Bottom) The interstellar probe should pass through the Sun's *heliopause* (the region where the Sun's solar wind hits the interstellar gas and dust and finally comes to a stop) in 2010. The interstellar precursor probe will continue sending back data until 2050, when it will have traveled a thousand AU from the sun. *(From Jaffe and Norton, in* Astronautics & Aeronautics *January 1980, p. 40. Copyright © American Institute of Aeronautics and Astronautics; reprinted with permission)*

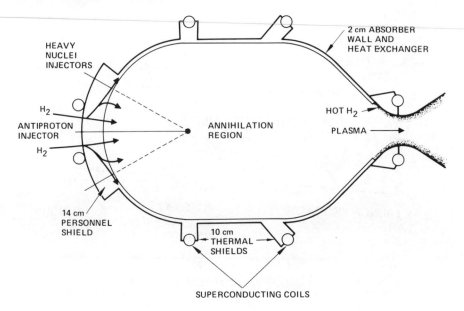

Figure 10–10.   The antiprotons annihilate with some of the hydrogen in the annihilation region of the combustion chamber, turning it into a hot hydrogen plasma. More hydrogen, flowing through a space between the chamber's inner and outer walls, cools the walls and protects the throat of the nozzle from the hot plasma.

the ride. The car passes the high school long before the skateboarder gets there, but the boy does finally coast up to the school's entrance.

JPL's Interstellar Precursor Mission, with its Pluto Orbiter tag-along, is going to take a long time to get a long way from anywhere. Suppose mirror matter is used for the propulsion system instead (Figure 10–10). Then it is possible to get anywhere in the solar system very quickly indeed.

Opening Up the Solar System

The important thing about mirror matter propulsion is that mass ratio is not a limiting factor. A space probe can be made as big and massive as necessary to carry whatever scientific instruments we'd like to send out there. All that is needed to get the necessary propulsion is to increase the amount of mirror matter used to react with the several tons of matter being carried for that purpose. Even advanced ion propulsion systems, such as those in Figures 10–11 and 10–12, would take years to reach the high velocity we'd need to completely escape the solar system. We can do it in days, if the spacecraft is hardy enough to take the stresses of massive acceleration, with mirror matter.

Some years ago science fiction writer Robert Heinlein doodled around with an idea for getting around the solar system quickly—very quickly. It was part of an article entitled "Where To?" appearing in Heinlein's book *Expanded Universe*, a collection of stories and essays published in 1980. We use his scenario with his permission. So—let's suppose it's the year 2006. Pluto and Charon (that's Pluto's moon, which is nearly as big as tiny Pluto itself) are about 32 AU from Earth, or 4.7 billion kilometers distant. Our mirror matter-powered space probe moves out of Earth parking orbit, accelerating on a tongue of white-hot flame at 9.8 meters per second per second—that's an acceleration of one G, the same as the force you and I feel standing on the surface of good old Terra Firma.

Question: How quickly can our Pluto probe get to Pluto? Answer: It depends on how long we let the mirror matter engines operate, because that will determine our speed. In fact, the math for this is so easy that even someone with "math-phobia" should have no trouble handling it. The

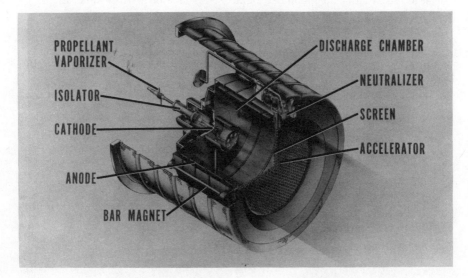

Figure 10–11 Ion space propulsion systems, which provide small but precise thrusts for very long periods of time, could power future space probes to the outer planets, including Pluto. (*NASA illustration*)

Figure 10-12   A solar-electric ion engine propels a space probe as it nears a comet in this artist's conception of a future NASA mission. *(NASA illustration)*

instantaneous velocity V is simply the product of acceleration and the elapsed time of acceleration:

$$V = at$$

The average speed at a constant acceleration is just half the maximum velocity at turnover.

$$s = \frac{V}{2} = \frac{at}{2}$$

The distance anything travels is equal to its speed times the amount of time it spends moving at that speed:

$$d = st$$

The probe is accelerating at a speed of 9.8 meters per second squared, one G. Suppose the engines keep operating until the probe is halfway to Pluto—2.35 billion kilometers. Then the engines shut off, the probe flips over, and the engines start again, to decelerate the spacecraft during the second part of the journey. The total time required is 1,385,052 seconds—about 385 hours, or sixteen days. *From Earth to Pluto in a little over two weeks!* Impossible? Not with mirror matter propulsion. The mathematics are simple, solid, and undeniable.

This simple exercise also tells us some other things. For example, the mirror matter propulsion system needn't run at a continual acceleration of 9.8 meters per second squared. If the continual acceleration and deceleration were only one-tenth of a G, or 0.98 meters per second squared, the probe would make it to Pluto in about 50 days. At one-hundredth of a G acceleration, this mirror matter-powered space probe would get to Pluto in about 23 weeks, or 160 days. That's still less than half a year, which is about the time a space probe would take today to get to Venus.

Also, we needn't run the mirror matter propulsion system in this "constant boost" manner. Suppose the probe accelerates at 1 G for just one hour instead of continually. Its velocity at the end of that hour will be an astounding 35.2 km/s. In that one-hour period the probe would travel about 63,500 kilometers, almost a fifth the distance from the Earth to the Moon. Suppose the probe now continued on to Pluto, engines off, coasting at 35 km/s. It will travel some three million kilometers per day if nothing slows it down. Of course, the gravitational pull of the Earth, Moon, Sun, and the other planets will slow it down. But for this "back-of-the-envelope" exercise we will ignore all that. At 35 km/s and three million kilometers per day, the Pluto probe will get to Pluto in about 1,550 days, or four years and three months. That's about how long it took for Voyager 2 to get from Earth to Saturn.

Sixteen days to Pluto is fantastic. Fifty days, or even six months, is wonderful. But even four years to Pluto isn't all that bad. At the very least, it's a timespan we are already used to. And for a robot space probe that will certainly be as hardy as the antique Voyager 2, four years should be a piece of cake.

Let's put these elapsed times in perspective. It took two years for Magellan's crew to circle the globe in a sailing ship. But that did not lead directly to a new wave of global migration. The Mayflower took ten weeks to get from England to America. And as long as that trip seemed to the passengers, it was still short enough to signal the real beginning of the European migration to the New World. Russian cosmonauts have toughed it out on the Salyut and Mir space stations for nearly a year. Their pioneering work in space will lead to the first human expeditions to Mars in the early twenty-first century. Those first Mars expeditions, though, will take as much time as did Magellan's global voyage. Human colonization of Mars will not yet be a reality.

However, the Russian experience with long-term space habitation in Mir and Salyut, and later American experience in the U.S. space station, will also lead to the return of humans to the Moon at the end of this century. Even with current chemical space propulsion systems, the Moon is just a three-day journey away from Tyuratam or Kennedy Space Center. It is safe to say, we think, that permanent human colonization of the Moon will begin in earnest in the first years of the twenty-first century. That will take place with or without mirror matter.

And further down the line? We will certainly have mastered the art and engineering of mirror matter space propulsion in the next 25 to 50 years. Pluto will be somewhat further from us in 2013 or 2038 than in the year 2006, but it won't be beyond our grasp. By then men and women will be used to living in space for years at a time. By then we will have a space propulsion system that can get objects and people from the Earth to the Moon in hours, and to Mars in weeks or even days. Robert Heinlein has already worked it out for us. Suppose Mars is in opposition to Earth, both planets are on the same side of the Sun, and Sun, Earth, and Mars make a straight line (this happens every 26 months or so). The average distance from Earth to Mars at opposition is about 60 million miles, or 96 million kilometers. Our United Spaceliner travels at a constant acceleration of one-tenth G, 98 centimeters per second per second. You could get from Earth to Mars in—hold on to your hat—*one week*. Round trip—with no stops for sightseeing if you bought the el cheapo fare—is about 15 days.

What if you're in a hurry, though? There's a political crisis boiling up. Dissident factions on Mars are threatening to take the colony out of its alliance with the North American government. Both the dissidents and the main political leaders have agreed to try and talk it out. They want you, Earth's best-known political troubleshooter, to mediate the dispute. How fast can you get there? Fast enough. At a constant one G boost (9.8 meters per second per second, constantly) for acceleration and deceleration, you'll be on the ground at Burroughs City a little more than two days after you leave your home in the Dyson City o'neill colony.

Now—suppose something *really* serious has happened. An unexpected meteor impact on Mars has severely disrupted the atmosphere. An extraordinarily powerful dust storm has destroyed three of the domes covering the martian colony's farm crops. What's left is not enough to feed all the people on Mars. If someone doesn't get enough food to Mars very quickly, a lot of people are going to starve to death. The Earth government calls on you, their ace troubleshooter, to come to the rescue.

The techs in the farm satellites for Dyson City and Feynman City quickly pull together a sufficient amount of foodstuffs to keep the Mars colony alive for a few weeks. That will give them breathing (actually, eating) room to rebuild the domes and plant some fast-growing crops. It will also give Earth- and o'neill-based farm-techs time to put together a long-term support plan until the colony is self-sufficient in food again.

How fast can you get the emergency foodstuffs to Mars? With your rescue ships powered by mirror matter, very fast. You boost to Mars at a constant 2 Gs—constant acceleration and then deceleration at twice the force of Earth gravity. But you and your handpicked crew can take it. In any case, you don't have to take it for too long—with a constant boost and de-boost of

2 Gs, you're going to land at Burroughs City just 39 hours after you leave Earth orbit! We're talking fast here.

If we can send robot space probes on a six-month journey to distant Pluto, then we can send ourselves to Pluto. And if our descendants in the year 2037 are anything like us today, they will go. That will signal the beginning of the human exploration of the entire solar system.

More than that: Any space propulsion system that can get us from Earth to Mars in two days, can get us from Earth to Pluto in two weeks. At that point, the solar system is not only opened up to human exploration, but also to human colonization.

Using mirror matter.

The next stop is the stars.

Turn Left at the Moon: VII

For the next two weeks the tour guides kept the vacationers busy visiting one site after another. An antimatter-powered hopper rocket zoomed them in one large leap first to the top of Mount Pavonis and then Mount Olympus, each some 27 kilometers high, and then back to Tharsis Base. Later an antimatter-powered crawler took them on an overnight, overland journey to the chaotic terrain that formed the western end of Mariner Valley.

"It looks like underground water subsidence to me," said Ginny. "But where did all the water go?"

Their next trip was aboard a six-person airglider. Although the martian atmosphere was very thin, an airplane could fly if it had a large enough wingspan, and not much mass. Fortunately, the fuel for the glider (which, stricly speaking, was not a glider at all, like the ones Eric flew back home, but had jets) weighed practically nothing at all . . . antimatter. The glider took them slowly down the length of Mariner Valley, often flying a kilometer or so below the tops of the three-kilometer-deep fissure in the skin of the planet. The canyon stretched on for thousands of kilometers. The glider stopped for the night at Chasma Corners, a luxury hotel located where the main chasm branched out into two smaller valleys.

The next day they took the northern route along Capri Chasm for a thousand kilometers, and then headed north across the Chryse Plains to visit the Thomas Mutch Memorial. The Viking 1 Lander still sat there, surrounded now by a low fence. The lander was covered with red dust, and beginning to look like the boulders among which it squatted.

Their next hop was a long one, as they flew directly north. They finally landed just after midnight in full sunlight at Boreal Camp, at the entrance to Boreal Canyon. Most of the personnel there were engineers making plans for an automated ice mine, and a water purification and piping station to supply the northern hemisphere. It would be ten times the size of the aqueduct system supplying the Republic of California.

"Oh, man, this is like some science fiction story," said Eric, looking at an artist's sketch of the system that hung in the camp's lobby. "Like maybe Bradbury's stories or something. Rocky. Where are the Martians? Lead me to 'em!"

The next day they visited the north polar ice cap. They traveled in a heated antimatter-powered snowmobile. It took them up the Boreal Canyon to see the layered cliffs of ice and dust that rose up hundreds of meters on each side. The snowmobile then headed down a small side-canyon for a few hundred meters and stopped.

No one said anything for long minutes. They all just stared at what was sticking out of the eroding ice cliff beyond. As they headed back to the camp the tour guide remarked, "Not Martians. It's pretty clear life never developed on Mars. We have found no trace of it, living or fossilized, in all our years of exploring.

"Whoever or whatever built that structure came from somewhere else."

Something suddenly crystallized inside Eric. It was so striking, so intense, that he could only close his eyes and hold his breath. He could feel tears coming. He finally turned to Ginny and whispered, "I'm going to find them.

"I swear to you. I swear to you. I'm going to find them."

Chapter Eleven

To the Stars

Interstellar missions will be many times more difficult to carry out than travel in the solar system. They won't be impossible, of course. But a dedicated interstellar space mission, even one with a robot space probe, is still many years, perhaps decades, from reality. There are other ways of interstellar travel besides chemical propulsion. One way may be mirror matter.

Distance and Time

It was only a few centuries ago that the human race realized those bright lights in the night sky were suns, like our Sun. People then realized that those other suns probably had worlds orbiting around them, some hopefully like our world. Since that time, one of the dreams of the human race has been to visit those other worlds in ships that travel between the stars. But as we began to realize the immensity of the distances that separate our star from the other stars, we despaired of ever building a starship using our puny technology. Now there is hope once again. We have found new sources of energy and produced new ideas for making starships. This has made it possible for the human race to start planning for a trip to the stars instead of just wishing for it.

It is difficult to comprehend the distances involved in interstellar travel. Of the billions of people living today on this globe, many have never traveled more than 40 kilometers (25 miles) from their place of birth. Of these billions, a few dozen have traveled to the Moon, which at a distance of almost 400,000 kilometers (250,000 miles), is ten thousand times farther away. Soon, one of our interplanetary space probes will be passing the orbit of Neptune, ten thousand times farther out at 4,000,000,000 kilometers (2,500,000,000 miles). However, the nearest star is at a distance of 40,000,000,000,000 kilometers (or 25,000,000,000,000 miles), ten thousand times farther than Neptune. The spacing between stars is so wide that even in terms of the distance between the Earth and the Sun (one astronomical unit or 1 AU),

215

the nearest star is 270,000 AU away. To cut interstellar distances down to verbal size, we use the *light-year*.

A light-year is the distance that light, traveling at 300,000 kilometers (186,000 miles) per second, travels in one year. From Earth, light takes about 1 second to reach the Moon, 8 minutes to reach the Sun, 4 hours to reach Neptune, and over 4 years to reach the nearest star system.

The nearest star system is called Alpha Centauri. Also known as "Rigil Kent," it is the brightest object in the southern constellation Centaurus, and the third brightest object in the sky after Sirius and Canopus. Alpha Centauri is not a single star, but a system of three stars. The nearest of those stars is a small red dwarf called Proxima Centauri, 4.3 light-years distant. The other two stars are one-tenth of a light-year farther and are called Alpha Centauri A and B. Alpha Centauri A is similar to our Sun, while B is slightly redder. These two stars orbit around each other every 80 years, while Proxima circles Alpha Centauri A and B with a period of millions of years. If our solar system had another sun in the place where Uranus is today, it would resemble the Alpha Centauri A and B system.

Some scientists think there may be another star in our solar system, in fact. The leading proponent of this hypothesis is University of California astronomer Richard Muller. He calls this star "Nemesis," and thinks it is responsible for the death of the dinosaurs 64 million years ago. Nemesis, says Muller, is in a distant orbit around the Sun that takes it through the so-called Oort Cloud of protocomets that surrounds our system. Its gravity dislodges protocomets, which every once in a great while come flashing down into the inner solar system. Some of them hit the Earth, kick up vast clouds of dust and gas, and create a mini-"nuclear winter" without the "nuclear." The drastic change in climate and temperature has killed many species of life, including the dinosaurs. Nemesis is so distant it takes it millions of years to make one revolution of the Sun. If this is in fact a true picture of things, if Nemesis does exist, it would be our equivalent of Proxima Centauri. The truth is, though, that there is still no evidence that Nemesis exists. Muller's Nemesis Hypothesis is still an hypothesis, not a fact.

Farther away in the heavens are some single star systems with stars that are also similar to our Sun. The planetary systems of these stars may be the most likely places to find an Earth-like planet. They include Epsilon Eridani at 10.8 light-years and Tau Ceti at 11.8 light-years. Within 12 light-years of the solar system there are 18 star systems with 26 stars (Table 11–1). Some of them may have planets. Canadian astronomer Bruce Campbell and two colleagues have found evidence, in fact, that Epsilon Eridani has a planetary companion some one to ten times the size of Jupiter. They also found evidence that four or five other stars within 50 light-years of the Sun also have planets. Other astronomers have found evidence that newly forming stars may frequently have the makings of planets about them. Steven Strom at the University of Massachusetts at Amherst and two colleagues studied

Table 11-1   The Nearest Stars or Star Systems and Their Distance
from the Solar System

Star	Distance (light-years)
Proxima Centauri (C) and Alpha Centauri (A) (B)	4.3
Barnard's Star	6.0
Wolf 359	7.7
Luyten 726–8 (A) and UV Ceti (B)	7.9
Lalande 21185	8.2
Sirius (A) (B)	8.7
Ross 154	9.3
Ross 248	10.3
Epsilon Eridani	10.8
Ross 128	10.9
61 Cygni (A) (B)	11.1
Luyten 789–6	11.2
Procyon (A) (B)	11.3
Epsilon Indi	11.4
Sigma 2398 (A) (B)	11.6
Groombridge 34 (A) (B)	11.7
Tau Ceti	11.8
Lacaille 9352	11.9

newly forming stars in a region of interstellar space called the Taurus
molecular cloud. They found evidence that protoplanetary disks of gas and
dust surround 20 of them. Those disks are likely to form into planets within
the next several hundred million years. Strom thinks that as many as 15
percent of all new stars have such disks, and may go on to have their own
full-fledged planetary systems.

So the planets may be out there. And perhaps those planets have
life . . . and intelligent beings who wonder about our star. . . . However,
it will take us a while to go out there ourselves and find out. No matter how
fast we can make a starship go, we must resign ourselves to the fact that
interstellar travel will always take a long time. Even if we had a starship
that traveled at the speed of light, it would take over 4.3 years to travel to
the nearest star system, then another 4.3 years before a message (or the starship) returns. So why should we bother going to the stars if it is so difficult?
One reason exists that should be obvious to us all, a reason built into our
genes. But it is so selfish that we often try to ignore it.

Reasons to Leave

We must go to the stars to survive. For the survival of the human race it is necessary that the human race leave the Earth. *Homo sapiens* has survived quite nicely on the Earth for tens of thousands and perhaps millions of years. So did the dinosaurs. But the dinosaurs are now extinct. If the human race is to survive, some small portion of it must leave this big blue egg and travel somewhere else to start a new branch of the human race. The universe is not malevolent towards humanity, but it is indifferent. If the Earth finds itself in the path of another celestial object, that object is going to hit the Earth. A 20-kilometer-wide asteroid could smash into the Earth *right now*—and we could do nothing to stop it. Such a cosmic collision would spell the end of civilization as we know it. It could even spell the end of the human species, not to mention a large percentage of all other life on Earth. This could be what happened to the dinosaurs, and could happen to us. *Lucifer's Hammer*, by Larry Niven and Jerry Pournelle, is a science fiction novel that vividly shows the results for humanity of a similar event.

But do we have to go to the *stars*? Why not just an L-5 colony, or Mars—or any of the other three dozen or so large solid bodies in the solar system? There are two answers: supernovas, and *homo sapiens*. If a nearby star exploded into a supernova, the radiation unleashed would be deadly to all life in the entire solar system, and not just on Earth. The two nearest stars that astronomers consider likely candidates for supernovas are the red supergiant stars Antares, 420 light years away, and Betelgeuse, 586 light years distant. Both are probably far enough away to be serious discomforts only if they explode. But what if a nearby blue giant or supergiant goes supernova? Supernova 1987a was just such a star—and its explosion was a surprise to astronomers. What if it's Fomalhaut (23 light years), or Altair (16 light years), or even Sirius (9 light years)? Then we're in serious trouble. Supernova 1987a showed us that we still have a lot to learn about supernovas.

And if the cosmos doesn't try to do us in, there's always the possibility that *we* will try to do us in. The nuclear apocalypse has been a favorite science fiction scenario since 1945. The recent theoretical studies of a so-called "nuclear winter" following a small nuclear war have reinvigorated that branch of SF. David Brin's novel *The Postman* is a look at what our future might be if things were merely awful after a nuclear war. The first part of *Between the Strokes of Night*, by Charles Sheffield, is a terrifying look at a truly apocalyptic future—a full-blown, pull-out-the-stops nuclear holocaust. The premise of Sheffield's novel, by the way, is that humanity survives this madness. The species survives because some of us *have moved off-planet*. Eventually humanity spreads throughout the galaxy, then the universe—and lives on to the final Big Crunch.

Another reason for interstellar travel would be to find other intelligent life-forms. The few millenia that this squabbling band of naked apes called the human race has been studying the stars is but an instant in the cosmic scale of time. The few decades that we have been trying to communicate with other beings around those stars is but the merest blink in that instant of time. We are just developing the technology needed for conducting astronomy with instruments instead of the naked eye alone. We have some good ideas, but we are still far from the technology for interstellar travel. It seems highly probable that any other race with which we make contact will have a technology that is thousands or millions of years ahead of ours.

If we can build large radio telescopes that have the inherent capability to transmit and receive radio signals over hundreds of light-years distance, then those more advanced races will have larger radio dishes, more powerful transmitters, and more sensitive receivers that can communicate between galaxies. (If they have a truly advanced technology, they may also have neutrino, and/or gravitational, and/or who-knows-what-other communication systems.) Yet we have listened for radio signals from the stars and we hear nothing. As physicist Enrico Fermi said so succinctly many years ago, "Where is everybody?"

If we can imagine at least one form of interstellar transport, be it expensive and slow, then surely they will have many forms of interstellar transport and communications. Some methods will use physical principles that we are familiar with. Other methods will use machines that work by some form of "magic" that we will understand only millions of years from now. Surely, if we can dream about travel to the nearest stars, then they are exploring the galaxy.

But if that is so, then where is everybody?

This paradox of the absence of evidence for extraterrestrial beings is one of the major unsolved questions of science. Scientific journals and the technical press contain article after article about the paradox. It is not necessary that the aliens come to visit us in person. Their robotic probes could do the job equally well—but we see no evidence of robots, either.

Some writers cite the absence of extraterrestrial visitors as proof that intelligent life has only formed once, and like it or not, we are that lone example of "intelligence" in the universe. They feel that it takes a remarkably fortuitous accident to create life, especially intelligent life.

It may be that it is difficult for life to start. But it is strange that we find amino acids, the chemical building blocks of life, in meteorites and the dust of interstellar space. In fact, it could be argued that intelligent life almost developed twice before on this planet.

In the distant past the Earth had only primitive forms of life. Today's hydra is a surviving example of these primitive life forms. The hydra consists of a few hundred cells formed into a foot, a gut, some tendrils, and a few primitive reflexes. From the early hydras eventually evolved the vertebrates with interior stiffening (who developed from worms), and the mollusks, with exterior stiffening (clams). The human is an intelligent vertebrate. The octopus is an almost-intelligent mollusk. The octopus brain is a five-lobed ring around the food intake, while the human brain is a two-lobed lump on top of the food intake. The eyes of the two species look similar, but developed independently. True, the octopus is not yet smart enough to be called an intelligent being. However, given a world without people to hunt them, a long dry spell to put stress on them, and a few million years to evolve some more, there could be intelligent air-breathing octopods designing starships to travel the galaxy.

Another example of an almost-intelligent species is, amazingly, a dinosaur. Apparently a branch of the dinosaurs called dromaeosaurids was on its way to developing bigger brains. They were already upright, had hands, possibly binocular vision, and possibly were warm-blooded. They disappeared 65 million years ago with the rest of the dinosaurs. If they hadn't, then Harry Harrison's science fiction novel *West of Eden* might be reality. Intelligent dinosaurs might have ruled the world.

Not all intelligent life-forms will have radio, however. For example, life could evolve on an ocean-covered world to produce intelligent whale-like or octopus-like creatures. These beings could be highly advanced in music, mathematics, philosophy, hydrodynamics, acoustics, and biology, but they would have no technology based on fire or electricity. They would not come to us. To meet these intelligent beings we would have to travel to them.

Interstellar Transport

It is difficult to go to the stars. They are far away and the speed of light limits us to a slow crawl along the starlanes. Decades and centuries will pass before the stay-at-homes learn what the explorers have found. The energies required to launch a manned interstellar transport are enormous, for the mass to be accelerated is large and the speed must be high. Yet even these energies are not out of the question once technology is moved out into space. There the constantly flowing sunlight is a never-ending source of energy—over a kilowatt per square meter, a gigawatt per square kilometer.

The first travelers to the stars will be robotic probes. They will be small and won't require all the amenities like food, air, and water that humans seem to find necessary. The power levels to send the first flyby probes are within the present reach of the human race. In fact, four different space probes are already on their way to interstellar space—Pioneer 10, Pioneer 11, Voyager 1, and Voyager 2. However, they were never designed to be *interstellar* probes. They're going to interstellar space because they have to. Their flybys past Jupiter, Saturn, Uranus, and Neptune (in 1989 for Voyager 2) have accelerated them to speeds beyond solar system escape velocity. It would be possible, though, to have the first *dedicated* interstellar probe on the way before the present millenium is out. As we noted in the last chapter, JPL already has studied one mission: TAU, the Thousand Astronomical Unit space probe.

The fact remains, however, that our present chemical propulsion systems are completely inadequate for traveling to the stars in any reasonable time period. In order for a chemically powered spacecraft to attain an interstellar flight velocity of only one percent of the speed of light, it would have to carry a fuel load many times bigger than the Sun. Chemical fuels will not get us to the stars—except very slowly.

Besides, there is a fundamental problem with any interstellar mission that travels at speeds less than 10 percent the speed of light. For even as a Robert Heinlein-type worldship is launched onto its centuries-long journey, propulsion engineers back on Earth will be dreaming about more advanced propulsion systems that can make starships that travel faster than the ship that is leaving. Within 20 or 30 years, those advanced propulsion systems will be a reality, and in another 10 to 20 years, a faster starship will zip past the lumbering worldship, explore the new star system first, then set up a welcoming party for the worldship colonists as they are picked up and brought in by a second wave of fast starships. Thus, until we run out of ideas for new propulsion systems, it would seem that no interstellar mission should be started if the mission takes more than 100 years. Instead, the money should be spent on research to build a faster propulsion system.

To carry out even a one-way robotic probe mission to the nearest star, Alpha Centauri, in the lifetime of the humans that launched the probe will require a minimum velocity of 0.1 c (10 percent of the speed of light). At that speed it will take the probe 43 years to get there and 4.3 years for the information to get back to us. To reach Epsilon Eridani or Tau Ceti in a reasonable time will require probe velocities of 0.3 c. At this speed it will take nearly 40 years to get there, plus another 11 to 12 years for the information to return to Earth.

Yet, although we need to exceed 0.1 c to get to any star in a reasonable time, if we can attain a cruise velocity of 0.3 c, then there are 18 star systems with 26 visible stars and hundreds of planets within 12 light-years. This many stars and planets within reach at 0.3 c should keep us busy exploring while our engineers are working on even faster starship designs.

Starship Designs

We need not be limited to chemical fuels in our journeys through the solar system and into interstellar space. Chemical rockets are just one of many methods of space propulsion (Tables 11–2 and 11–3). Once we realize that, our imaginations are free to design starships that will turn science fiction into reality. What kinds of starships can we envision? It turns out there are many, each using a different technology: electric propulsion, fusion rockets, nuclear pulse rockets, interstellar fusion ramjets, ram-augmented fusion rockets, beamed power propulsion, and antimatter rockets.

Electric Ion Propulsion   The most advanced form of flight-tested propulsion we have today is *electric ion propulsion*. In this type of spacecraft some energy source is used to produce electricity and the electricity is used to ex-

Table 11-2 Fifty-two Space Propulsion Concepts

Concept	Energy Source	Reaction Material or Method
1 Fanjet	Air-Fuel Combustion	Atmosphere
2 Aerobrake	Kinetic Energy	Atmosphere
3 Turbo-Ramjet	Air-Fuel Combustion	Reaction Byproducts
4 Ramjet	Air-Fuel Combustion	Reaction Byproducts
5 Scramjet	Air-Fuel Combustion	Reaction Byproducts
6 Chemical Rocket	Fuel-Oxidizer Combustion	Reaction Byproducts
7 Scramjet Gun	Fuel-Oxidizer Combustion	Reaction Byproducts
8 Pure Fusion Rocket		
8.1 Magnetic Containment	Fusion Reactor	Reaction Byproducts
8.2 Inertial Confinement	Fusion Reactor	Reaction Byproducts
9 Pure Antimatter Rocket	Antiproton Annhilation	Reaction Byproducts
10 Pulsed Fission Rocket	Nuclear Explosion	Reaction Byproducts
11 On-Site Fuel Manufacture	Extraterrestrial Materials	Reaction Byproducts
12 Nuclear Rocket	Fission Reactor	Heated Hydrogen Gas
13 Fusion-Hydrogen Rocket	Fusion Reactor	Heated Hydrogen Gas
14 Antimatter-H_2 Rocket	Antiproton Annhilation	Heated Hydrogen Gas
15 Electric Rail Rocket	Electric Power Line (Rail)	Heated Hydrogen Gas
16 Microwave Rocket	Microwave Beam	Heated Hydrogen Gas
17 Laser Rocket	Laser Beam	Heated Hydrogen Gas
18 Solar-Thermal Rocket	Sunlight	Heated Hydrogen Gas
19 Railgun	Electric Power Line (Rail)	Plasma
20 Nuclear-Ion Rocket	Fission Reactor	Ion
21 Microwave-Ion Rocket	Microwave Beam	Ion
22 Laser-Ion Rocket	Laser Beam	Ion
23 Photovoltaic-Ion Rocket	Sunlight(Photovoltaic)	Ion
24 Photon Rocket	Antiproton Annhilation	Photons
25 Starwisp	Microwave Beam	Photons
26 Laser Lightsail	Laser Beam	Photons
27 Solar Lightsail	Sunlight	Photons
28 Mass Driver Launcher	Electric Power Line (Rail)	Magnetic Field
29 Mass Driver Engine	Sunlight(Photovoltaic)	Magnetic Field
30 Electrodynamic Engine	Sunlight(Photovoltaic)	Magnetic Field

Concept	Energy Source	Reaction Material or Method
31 Launch Loop	Kinetic Energy	Magnetic Field
32 Artillery	Fuel-Oxidizer Combustion	Gas Expansion in Tube
33 Rocket Fed Gas Gun	Fuel-Oxidizer Combustion	Gas Expansion in Tube
34 Nuclear Reactor Gas Gun	Fission Reactor	Gas Expansion in Tube
35 Nuclear Explosion Gas Gun	Nuclear Explosion	Gas Expansion in Tube
36 Discharge Heated Gas Gun	Electric Power Line	Gas Expansion in Tube
37 Compressed Gas Gun	Compressed Gas	Gas Expansion in Tube
38 Underwater Gas Gun	Compressed Gas	Gas Expansion in Tube
39 Piston Driven Gas Gun	Kinetic Energy	Gas Expansion in Tube
40 Leveraged Catapult	Potential Energy	Mechanical
41 Towers	Potential Energy	Mechanical
42 Tethers		
42.1 Terrestrial	Kinetic Energy	Mechanical
42.2 Orbital	Kinetic Energy	Mechanical
43 Tow-Rope	Kinetic Energy	Mechanical
44 Dumb-Waiter	Extraterrestrial Materials	Mechanical
45 Flyby	Kinetic Energy	Gravity
46 Solar-Electric Arc-Jet	Sunlight (Photovoltaic)	Plasma
47 Air-Breathing Laser Propulsion	Laser Beam	Atmosphere
48 Alpha Particle Rocket	Nuclear Fission	Alpha Particles
49 Thermoeletcric Ion	Nuclear Fission	Ion
50 Buoyant Scramjet Gun	Fuel-Oxidizer Combustion	Reaction Byproducts
51 Hydrogen Tunnel	Kinetic Energy	Hydrogen Tunnel
52 Railgun Engine	Sunlight (Photovoltaic)	Plasma

Most of these concepts are still theoretical, but are scientifically possible. Three of them (9, 14, and 24) use mirror matter.

(Courtesy of Dani Eder)

Table 11-3 Theoretical Space Propulsion Concepts— Energy Source vs. Reaction Material or Method

Energy Source	Atmosphere	Reaction Byproducts	Heated Gas	Plasma	Ion	Photons	Magnetic Field	Gas Expansion In Tube	Mechanical	Gravity
Air-Fuel Combustion	(1) Fanjet	(3) Turbo-Ramjet, (4) Ramjet, (5) Scramjet								
Fuel-Oxidizer Combustion		(6) Chemical Rocket, (7) Scramjet Gun, (50) Balloon Gun						(32) Artillery, (33) Rocket Fed Gas Gun		
Nuclear Fission		(48) Alpha Particle Rocket			(49) Thermo-Electric Ion					
Fission Reactor			(12) Nuclear Rocket		(20) Nuclear Ion Rocket			(34) Particle Bed Reactor Gas Gun		
Fusion Reactor		(8) High Efficiency Fusion Rocket	(13) High Thrust Fusion Rocket							
Antimatter Annihilation		(9) High Efficiency Antimatter Rocket	(14) High Thrust Antimatter Rocket			(24) Photon Rocket				
Nuclear Explosion		(10) Orion Rocket						(35) Nuclear Pumped Gas Gun		
Electric Power Line (Rail)			(15) Rail Rocket	(19) Railgun			(28) Mass Driver Launcher	(36) Discharge Heated Gas Gun		
Microwave Beam			(16) Microwave Rocket		(21) Microwave Ion Rocket	(25) Starwisp				
Laser Beam	(47) Waverider		(17) Laser Rocket		(22) Laser Ion Rocket	(26) Laser Lightsail				
Sunlight (Photovoltaic)			(18) Solar Thermal Rocket	(46) Solar Electric Arc-Jet	(23) Photovoltaic Ion Rocket	(27) Solar Lightsail				
Compressed Gas							(29) Mass Driver Reaction Engine (30) Electrodynamic (52) Railgun Engine	(37) Gas Gun (38) Underwater Gas Gun		
Potential or Kinetic Energy	(2) Aerobrake (51) Hydrogen Tunnel						(31) Launch Loop	(39) Piston Driven Gas Gun	(40) Leveraged Catapult (41) Towers (42) Tethers (43) Tow Rope	(45) Planet or Satellite Flyby
Extraterrestrial Materials	(11) On-Site Fuel Manufacturing								(44) Dumb-Waiter	

(*Courtesy of Dani Eder*)

pel charged ions of an inert reaction mass such as mercury, cesium, xenon, or argon atoms at high speed to provide thrust. In most electric propulsion systems, the electrical energy is obtained from solar cells that convert sunlight into electricity. Solar electric propulsion, by the way, is not a true rocket system, since the vehicle does not carry its own energy source (although it does have to carry the not inconsiderable weight of the solar cells and power conditioners that convert the sunlight into electrical power). Unfortunately for interstellar missions, the sunlight rapidly becomes weaker as the spacecraft leaves the solar system. Solar electric propulsion will not get us to the stars.

An alternate energy source is a nuclear fission reactor used to supply electric power (see Figure 11–1). A nuclear electric rocket would be capable of reaching a velocity of 150 km/s after using up all the fuel in the nuclear reactor. This is a much higher velocity than has been reached by any space vehicle built to date. But to put it in another perspective, 150 km/s is only 1/2,000 of the speed of light. Such a spacecraft would be useful for exploring

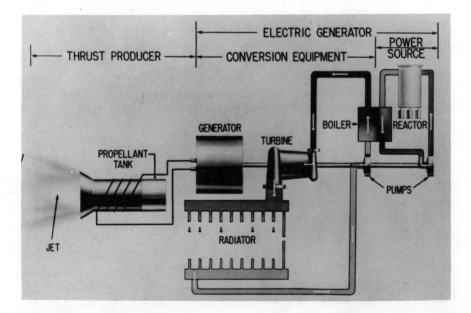

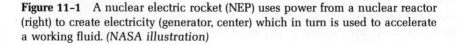

Figure 11-1 A nuclear electric rocket (NEP) uses power from a nuclear reactor (right) to create electricity (generator, center) which in turn is used to accelerate a working fluid. *(NASA illustration)*

extrasolar space to search for trans-Plutonian planets and nearby "brown dwarf" stars. A nuclear-powered "ion propulsion" drive is used by the U.S.S. Enterprise in the Star Trek TV shows, movies, and books. But as Captain Kirk has found out, although 150 km/s is useful for cruising near planets, at that speed the nearest star is 10,000 years away. Nuclear electric propulsion will not get us (or the Enterprise) to the stars.

Fusion Rockets   In scientific laboratories all around the world there are large, well-funded, intense research efforts to achieve controlled nuclear fusion of various types of hydrogen nuclei into helium nuclei. The result will be the release of large amounts of energy. To date, the research on controlled fusion has not even achieved the engineering "break-even" point, where the fusion reactor produces enough power to keep itself running. Research is continuing, though, and soon one of the many different approaches will work.

Once controlled fusion has been achieved in the laboratory, scientists can start designing an interstellar rocket based on those types of reactors that turn out to be feasible. If controlled fusion is achieved by compression and heating of a plasma in a magnetic bottle, then perhaps all that is needed to convert the reactor into a rocket is to allow the magnetic bottle to "leak" a little bit, and the hot plasma exhaust will produce thrust. If controlled fusion is achieved by implosion of micropellets with beams of laser light, electrons, ions, or high-speed pellets, then the same technique can be used to implode the pellets in the throat of a rocket nozzle made of magnetic fields. The exploding plasma will then be turned into directed thrust.

The reaction presently being used in both the containment and implosion fusion research projects involves the fusion of deuterium (a hydrogen nucleus with one extra neutron) and tritium (a hydrogen nucleus with two extra neutrons). Tritium is radioactive, with a half-life of 12.3 years, Thus interstellar spacecraft using the D-T reaction must have a method of generating tritium on board. Alternatively, additional work on containment and implosion fusion techniques could produce the pressures, temperatures, and densities needed to achieve fusion with deuterium and helium-3 (D-He^3), D-D, or proton-proton reactions.

Depending upon the success of the research on ground-based fusion power plants, we may someday have a fusion rocket engine. Unfortunately, most fusion machine designs being considered are very massive for the amount of energy they produce and are unlikely to provide rockets with the thrust-to-weight ratio needed for achieving fractions of the speed of light in a reasonable acceleration period. Fusion rockets *can* get us to the stars—if they aren't too heavy.

Nuclear Pulse Propulsion   Although we do not yet have controlled fusion, we are experts on *uncontrolled* fusion—thermonuclear bombs. One of the oldest designs for an interstellar vehicle is one that is propelled by such bombs. The first design was the Orion spacecraft (Figure 11–2). Orion was

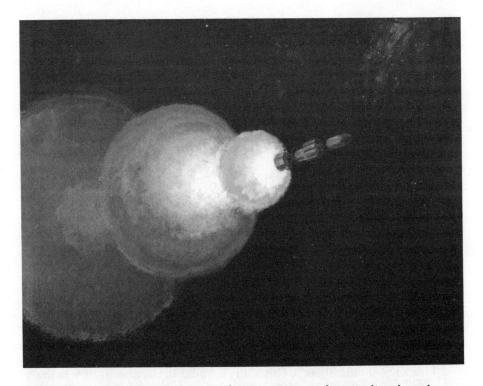

Figure 11-2 This artist's conception shows an Orion rocket traveling through space, propelled by small nuclear bombs exploding behind it. Debris from the explosions hits a pusher plate at the rear of Orion and the shock waves push the vehicle through space. Orion, originally designed by Stanislaw Ulam and refined by Theodore Taylor and Freeman Dyson, was never built. *(Artist: Seichi Kiyohara/Hughes Research Laboratories)*

invented in the late 1950s at Los Alamos National Laboratory by an inventor of the hydrogen bomb, Stanislaw Ulam. The original goal of the Orion Project was to send a manned spacecraft to Mars and Venus by 1968 at a fraction of the cost of the Apollo project.

The Orion vehicle works by ejecting a small nuclear bomb out the rear of the vehicle through a small hole in a "pusher plate," where the bomblet explodes. The debris from the nuclear explosion strikes the pusher plate, which absorbs the impulse from the explosion and transfers it through large "shock absorbers" to the main spacecraft. Although it seems amazing that anything could survive a few dozen meters away from a nuclear explosion, a properly designed pusher plate with a surface like a re-entry heat shield can stand not one, but many thousands of such nuclear explosions.

Physicist Freeman Dyson took these well-engineered ideas for an interplanetary spacecraft and extrapolated them to an interstellar spacecraft. The ship would necessarily be large, with a payload of some 20,000 tons (enough to support a small township of many hundred crewmembers). The total mass would be 400,000 tons, including a fuel supply of 300,000 nuclear bombs weighing about 1 ton each. (This is approximately the world's supply of nuclear bombs—what an excellent way of disposing of them!)

The bombs would be exploded once every three seconds, pushing the spacecraft at an acceleration equal to Earth's gravity. After 10 days the Orion starship would reach a velocity of 1/30 of the speed of light. At this speed the Orion spacecraft would reach Alpha Centauri in 130 years. It would take over 300 years to get to the interesting star systems like Tau Ceti and Epsilon Eridani. To give this ship a deceleration capability at the target star, it would need to be redesigned to have two stages, with the first stage weighing 1,600,000 tons.

These slow travel times for nuclear powered starships are longer than the present lifetime of a human being, so it looks like the travelers on our slow nuclear starships will have to be long-lasting robots rather than ephemeral humans, unless the generation worldship approach is used or some technique can be found that will allow a human crew to live longer than our presently allotted three-score-and-ten-year lifetimes. Nuclear pulse propulsion will get us to the stars—slowly. Of course, it *could* get us to Mars, and quickly, which was its original purpose. Orion was never built, however, because the Nuclear Test Ban Treaty, signed in 1963, prohibits the explosion of nuclear bombs in outer space.

In science fiction, though, Orion *has* been built. Once again, it was Larry Niven and his colleague Jerry Pournelle who did the trick. Their Orion ship wipes out Bellingham, Washington, but also provides a smashingly exciting climax to their novel *Footfall*.

Although the Orion spacecraft would have a minimal performance for a starship, it is one form of interstellar transport that could have been built and sent on its way in the last decade. The reason for the relatively poor performance of the Orion vehicle is that it uses the rocket principle, where the vehicle is required to carry its own propulsion energy and reaction mass along with it. Unless we are willing to consider enormous mass ratios, the terminal velocity achievable is roughly equal to the exhaust velocity. That in turn is proportional to the square root of the fraction of the mass of the fuel that's converted into energy. A fusion reaction converts less than 1 percent of its mass into energy, so the exhaust velocity is limited to 3 percent or so. If the rocket concept is to be used for rapid interstellar transport, a nuclear fuel will be needed that has a higher energy conversion efficiency than fusion. One such high-energy fuel is antimatter. Before we get to that, though, here are some other propulsion systems to consider.

Interstellar Fusion Ramjet   One of the oldest interstellar transport techniques is the Bussard *interstellar ramjet* invented back in 1960 by Robert W. Bussard, a research scientist in controlled fusion. The interstellar ramjet consists of a payload, a fusion reactor, and a large scoop to collect the hydrogen atoms in space (Figure 11–3). The interstellar ramjet carries no fuel

Figure 11-3   A Bussard ramjet travels through interstellar space by scooping
up hydrogen atoms for fuel with some kind of scoop. This kind of vehicle is
an example of a "rocketless rocket," since it does not carry its own fuel with it.
(Artist: Seichi Kiyohara/Hughes Research Laboratories)

because it uses the scoop to collect the hydrogen atoms that are known to
exist in space. The hydrogen atoms are used as fuel in the fusion reactor,
where the fusion energy is released. The reaction products (usually helium
atoms), provide the thrust for the vehicle.

Bussard originally estimated that a 1,000-ton vehicle would require
a frontal intake area of about 10,000 square kilometers to achieve an
acceleration of 1 G (9.8 m/s^2) through interstellar space with a density of
1,000 hydrogen atoms per cubic centimeter. That requires a scoop that is
100 kilometers or 60 miles in diameter. The ramjet "take-over" velocity is
extremely low, so that conventional chemical rockets could provide the

initial acceleration. As the vehicle increases its velocity into the relativistic region, the interstellar fuel flow appears to increase in density due to the Lorentz contraction in the vehicle's space-time. Thus, the faster the ramjet goes, the faster it *can* go.

If an interstellar ramjet could ever be built, it would have many advantages over other possible starships. Since it never runs out of fuel like fuel-carrying rockets, and never runs away from its source of power like an externally powered propulsion system, it can accelerate indefinitely. The interstellar ramjet is the only starship design known that can reach the relativistic velocities where ship time becomes many times slower than Earth-time. This would allow human crews to travel throughout the galaxy or even between galaxies in a single human lifetime.

A lot of invention and research is needed, however, before the Bussard interstellar ramjet becomes a reality. First controlled fusion must be achieved. The fusion reactor must not only be lightweight and long-lived, but must also be able to fuse protons, not the easier-to-ignite mixture of deuterium and tritium. The reactor must be able to fuse the incoming protons without slowing them down; otherwise the frictional loss of bringing the fuel to a halt, fusing it, and reaccelerating the reaction products will put an undesirable upper limit on the maximum velocity attainable.

Many astronomers currently think the solar system is not far from the edge of a warm, mostly neutral hydrogen gas cloud a few tens of light-years in radius with a relatively low density of 0.1 atoms per cubic centimeter (as compared to the 1,000 atoms per cubic centimeter assumed by Bussard). This region is surrounded by a pervasive hot, very low-density ionized plasma with a density less than 0.01 ions per cubic centimeter that extends some 150 light-years or more in all directions observed. This ubiquitous low density hot gas "bubble" is most likely the result of past supernova explosions—not promising for would-be Bussard ramjet users.

The major difficulty, though, with the ramjet starship is the design of the scoop, which must be ultra-large and ultra-light. If the interstellar hydrogen were ionized, then a large, super-strong magnet might be sufficient to scoop up the charged protons. Although some distant stars have clouds of ionized hydrogen near them, most of the hydrogen near the solar system is neutral. Schemes for ionizing the hydrogen have been proposed, but they are neither light in weight nor low in power consumption. For now, in regions near the solar system, the interstellar ramjet remains in the category of science fiction.

The concept of picking up your fuel along the way as you journey through "empty" space is too valuable to be discarded easily, however. Perhaps future scientists and engineers will keep working away on the remaining problems until this beautiful concept is turned into a real starship.

The Bussard interstellar ramjet is an early example of what we call "rocketless rocketry": techniques for traveling though space without being bound by all the limitations of a standard rocket.

Rocketless Rocketry

It is not necessary to use the rocket principle to build a starship. If we examine the components of a generic rocket, we find that it consists of payload, structure, reaction mass, energy source, a reactor to put the energy in the reaction mass, and a thruster to expel the reaction mass to provide thrust. In most rockets the reaction mass and energy source are combined together into the chemical fuel. The fuel is burned in the engine, which is a reaction chamber or reactor. The hot reaction products come streaming out the rocket nozzle or thruster. Because a standard rocket has to carry its fuel along with it, its performance is significantly limited. For missions where the final vehicle velocity is much greater than the exhaust velocity, the amount of fuel needed rises very quickly to impossible amounts.

It is possible to conceive of space vehicle designs that do not use the rocket principle and thereby avoid the exponential mass growth implicit in the design of a standard rocket. These rocketless rockets are excellent candidates for starships. The Bussard interstellar ramjet is one example. It carries no fuel, but scoops up its reaction mass and energy source from "empty" space. Some other rocketless rockets are described on the pages following.

Ram-Augmented Interstellar Rocket   There are some versions of the Bussard interstellar fusion ramjet that might become feasible sooner than the pure fusion ramjet. These use combinations of rocket technology, ram-scoop technology, and beamed power technologies to produce an interstellar vehicle system. As we just saw, the Bussard does not carry its reaction mass along with it. Instead, it scoops up the hydrogen atoms in interstellar space and uses them as the reaction mass. In the straight rocket mode, the energy source is on board and is either a nuclear power plant or stored antimatter. The energy is used to heat the hydrogen reaction mass, which is then exhausted from the vehicle to provide thrust. An alternate energy source could be in the form of beamed power (laser beams or neutral particle beams of antimatter) from the solar system. The beamed power would be used to heat the collected interstellar hydrogen to provide thrust. This *beam-energized ram-augmented* interstellar vehicle would not have to carry its energy source or reaction mass, but it still would have to carry its reactor and thruster, as well as its payload and structure.

The ram-augmented interstellar rocket, despite its potential advantages, has the same fundamental problem as the Bussard interstellar ramjet. There are no physically feasible designs presently known for constructing a lightweight scoop that can collect enough hydrogen reaction mass from the low-density, mostly neutral interstellar medium surrounding the solar system to achieve reasonable accelerations and reasonable cruise velocities. New ideas for scoops are badly needed. Until these new ideas are generated, any propulsion system that depends upon scooping up the interstellar medium for fuel is not presently feasible. Ramscoop vehicles will not get us to the stars—yet.

Laser-Pushed Lightsails   We just mentioned a power source that makes possible another whole class of spacecraft. These do not have to carry along any energy source or reaction mass or even a reactor. They are spacecraft that work by *beamed power propulsion* (Figure 11–4). They consist only of payload, structure, and thruster. In a beamed power propulsion system, the heavy parts of a rocket (the reaction mass, the energy source, and the reactor that takes the energy from the energy source and puts it into the reaction mass) are all kept in the solar system. Here, around the Sun, there are unlimited amounts of reaction mass readily available, and the energy source (usually the abundant sunlight) and reactor can be maintained and even upgraded as the mission proceeds.

Figure 11-4 High-powered lasers may someday be used to provide energy for space propulsion close to Earth. This illustration shows an Orbital Transfer Vehicle receiving energy from a continuous-wave laser beamed from a solar-powered satellite in geosynchronous orbit. The laser beam is focused down to a point and enters the OTV through a transparent window. The energy heats . the OTV's fuel (or working fluid) to a hot gas or plasma, which streams out the rocket nozzle for thrust. (*Courtesy NASA/Marshall Space Flight Center*)

The best of the beamed power propulsion systems uses large sails of light-reflecting material pushed by the photon pressure from a large laser array in orbit around the Sun. In laser-pushed lightsail propulsion, light from a powerful laser is bounced off a large reflective sail surrounding the payload. The light pressure from the laser light pushes the sail and payload, providing the needed thrust. The sails that the laser craft would use would be advanced versions of the sun-pushed lightsails that have been studied by the NASA Jet Propulsion Laboratory for comet missions and fast trips to the asteroid belt (Figure 11-5). To travel out between the stars where the solar sunlight gets dim, though, it would be necessary to use a laser instead of sunlight.

The laser would be an advanced version of the high-power laser arrays presently being studied for the SDI of the Department of Defense. The important thing to realize is that no scientific breakthroughs are needed to build this starship. The basic physical principles of the lasers, the transmitter lens, and the sail are known. All that is required to make the laser sail starship a reality is a lot of engineering (and a lot of money).

Figure 11-5   Light sails have already been studied by NASA for use with space probes. This "heliogyro" concept (so named because its sails look like helicopter blades) was one proposal for a Halley's Comet probe. Each sail is over seven kilometers (4.3 miles) long. *(NASA illustration)*

The basic technique for the laser-pushed lightsail is one of the few scientific ideas that appeared first in science fiction, then later in a scientific publication (the references in the scientific paper give credit to the novel as the first source of the concept). Using various versions of the basic technique described in the novel, a manned spacecraft can be built that not only can travel at reasonable speeds to the nearest stars, but can also stop, then return its crew back to Earth again within their lifetime. It will be some time before our engineering capabilities in space will be up to building the laser system needed, but there is no new physics involved, just a large-scale engineering extrapolation of known technologies.

The lasers would be in space and energized by sunlight collected by large lightweight reflectors made of thin aluminum-coated plastic. The lasers themselves would be best in orbit around Mercury. The sunlight there is more intense, and the gravitational pull of Mercury would keep the lasers from being "blown" away by the reaction from their light beams. They would use the abundant sunlight at Mercury's orbit to produce coherent laser light, which would then be combined into a single coherent laser beam and sent out to a transmitter lens floating between Saturn and Uranus.

The transmitter lens would be a segmented ring lens consisting of wide rings of extremely thin plastic film alternating with empty rings. The width of the rings would be tuned to the laser frequency. The transmitter lens would not be in orbit around the Sun, but would either be freely falling (very slowly at that distance from the Sun), or "levitated" in place by rockets or by the momentum push from a portion of the laser light passing through it. The segmented ring lens would be 1,000 kilometers (600 miles) in diameter and weigh about 560,000 tons. A lens this size can send a beam of laser light over 40 light-years before the beam starts to spread.

We will want a starship design that can carry out roundtrip missions to stars as distant as Tau Ceti and Epsilon Eridani within a human lifetime. This will require a lightsail that is the same size as the transmitting lens. The lightsail would be built in three sections. There would be an inner payload sail that is 100 kilometers (60 miles) in diameter. Surrounding that would be an inner ring-shaped sail that is 320 kilometers (200 miles) in diameter with a 100-kilometer hole. Surrounding that would be an outer ring-shaped sail that is 1,000 kilometers (600 miles) in diameter with a 320-kilometer diameter hole. The total structure would mass 80,000 tons, including 3,000 tons of payload consisting of the crew, their habitat, their supplies, and their exploration vehicles.

The entire lightsail structure would be accelerated at 30 percent of Earth gravity by 43,000 terawatts of laser power. (Since the Earth only produces about 1 terawatt of electrical power, we would certainly want to use the free solar power in space instead of trying to get our power from Earth.) At this acceleration, the lightsail would reach a velocity of half the speed of light in 1.6 years (Figure 11–6). The expedition would reach Epsilon Eridani in 20 years Earth time and 17 years crew time.

Figure 11-6 Lasers or microwave beams may be combined with sails to power interstellar probes. Here a superpowerful laser back at Earth accelerates a crewed interstellar probe on its way to another star. The vehicle is mostly a gigantic "light sail." The laser beam appears "red-shifted" because the crew is moving away from it. The laser accelerates the vehicle to half the speed of light. It takes it twenty years Earth time to get to the star Epsilon Eridani, about eleven light-years from Earth. (*Artist: Seichi Kiyohara/Hughes Research Laboratories*)

At 0.4 light-years from the target star, the 320-kilometer rendezvous portion of the sail would be detached from the center of the lightsail and turned to face the large ring sail that remains. The inner portions would be allowed to lag behind while they are being turned to face the large outer ring sail. The laser light coming from the solar system would reflect from the outer ring sail acting as a retro-directive mirror. The reflected light would decelerate the two inner portions and bring them to a halt at Epsilon Eridani (Figure

11–7). After the crew explores the system for a few years (using their lightsail as a solar sail), it would be time to bring them back.

To do this, the 100-kilometer diameter return sail would be separated out from the center of the 320-kilometer rendezvous sail, and they would be turned to face each other. If someone back in the solar system remembered to turn on the laser beam 12 years earlier, the laser beam from the solar system would hit the ring-shaped remainder of the rendezvous sail and be reflected back on the return sail, sending it on its way back to the solar system. As the return sail approaches the solar system 20 Earth-years later, it would be brought to a halt by a final burst of laser power (Figure 11–8). The members

Figure 11-7   The inner part of the sail detaches from the outer ring and is turned to face it. The laser beamed from Earth strikes the ring and bounces back to hit the decelerator sail, slowing it to a stop in the Epsilon Eridani system. The crew spends five years exploring, using their sail as a solar sail to "tack" around in the system. *(Artist: Seichi Kiyohara/Hughes Research Laboratories)*

of the crew will have been away 51 years (including 5 years of exploring), have aged 46 years, and be ready to retire, write their memoirs, and hit the Tri-V talkshow circuit.

The major problem with the laser-pushed lightsail is that it is extremely energy-inefficient. In any propulsion system, one wants as much of the energy in the fuel as possible to end up as kinetic energy in the vehicle. That means tailoring the exhaust velocity of the reaction mass so that it leaves the vehicle at the same speed that the vehicle is going. The reaction mass is then left motionless in space, with no kinetic energy left in it. All of the propulsive energy has gone into the space vehicle.

In a laser-pushed lightsail, the laser photons bounce off the sail, giving it

Figure 11-8 Many years later, the crew is returning to Earth. The laser (now blue-shifted because the probe is approaching the light) fires again to slow the return portion of the light sail and its passengers to a stop. They have been gone for 51 years and have aged 46 years. (Artist: Seichi Kiyohara/Hughes Research Laboratories)

a push, and reflect backward, still containing most of their initial energy. If the lightsail is moving greater than half the speed of light, the reflected laser light will be red-shifted by the doppler effect, indicating that it deposited a lot of energy in the sail. But most near-term interstellar missions will be at velocities less than half the speed of light. In this case, the laser photon is only slightly red-shifted, indicating that only a small portion of the initial photon energy got put into the sail. As a result, very high levels of laser power have to be generated back in the solar system in order to put just a small amount of energy into the sail. The process is very energy-inefficient, and therefore costly in terms of the size of the facilities that are needed to process the sunlight, even if the sunlight is free. Laser-pushed lightsails can get us to the stars—but they are expensive.

What, then, would be a way to get to the stars that is efficient, economical, and powerful? The most efficient, most powerful, nearest term, and fastest interstellar propulsion system known is the antimatter rocket.

Antiproton Annihilation Propulsion

Mirror matter represents a highly concentrated form of energy with the ability to release "200 percent" of its rest mass as energy when it annihilates with an equal amount of normal matter. Mirror matter annihilation is a thousand times more powerful than nuclear fission and a hundred times more powerful than nuclear fusion. A spacecraft that uses mirror matter as its source of propulsion energy could "drive" anywhere in the solar system with mission times ranging from days to weeks, and could even travel to the nearest stars in a small fraction of a human lifetime (Figure 11–9).

As we saw earlier, one of the amazing things about antimatter-powered rockets is that if they are designed to minimize the amount of antimatter needed, all antimatter-rocket-powered vehicles have the same maximum fuel-to-vehicle ratio. The maximum mass ratio of the fully fueled spacecraft to the empty spacecraft turns out to be about 5, regardless of whether they're flying from low Earth orbit (LEO) to geosynchronous Earth orbit (GEO), or to the distant stars. The ratio is also independent of the efficiency and mass of the rocket engine. It just takes longer to get there at low efficiency or with heavier engines. For missions that require characteristic velocities less than half the speed of light, equal amounts of matter and antimatter should not be used. Instead, the energy from a small amount of antimatter should be used to heat a much larger amount of reaction mass.

This is definitely not the case for chemical rockets. An easy mission like sending a communication satellite from LEO to GEO can be done with a small solid rocket built into the satellite with a mass ratio of 3. To send that same payload to Mars, one would need a fuel tank as big as the space shuttle main tank with a mass ratio of 25. To send that same payload to

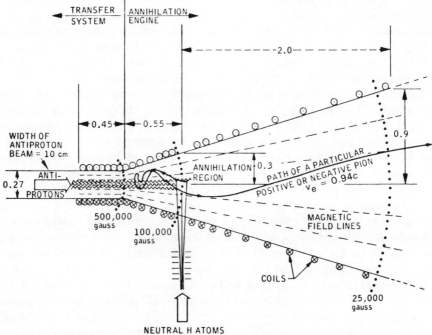

NEUTRAL H ATOMS

Figure 11-9 This is a diagram of a Morgan antimatter space propulsion engine, devised by David Morgan of the Lawrence Livermore National Laboratory. A beam of antiprotons annihilates with a beam of neutral hydrogen atoms at the top of the rocket engine (left). Superpowerful magnetic coils force the pion particle fragments of the annihilation to go streaming out the nozzle of the rocket. The exhaust velocity is 94 percent the speed of light! The nozzle is about three meters long.

the rings of Saturn would require a fuel tank as big as an oil tanker with an "impossible" mass ratio of 20,000. In contrast, with mirror matter rockets the mass ratio is never greater than 5, and often less.

So—we've got a large pile of mirror matter stored away. Let's go to the stars.

Ad Astra! *To the Stars!*

Suppose we wish to send a one-ton scientific spacecraft at one-tenth the speed of light on a flyby of Alpha Centauri, 4.3 light-years away. To carry out this mission using an antimatter rocket would require a small fuel tank holding a mere four tons of some working fluid like liquid hydrogen, methane, or water, and another, much smaller fuel tank of negligible mass, holding 18 kilograms (40 pounds) of antimatter. The total mass of the vehicle is essentially five tons per each one ton of payload, or a mass ratio of 5. At 10 percent of light speed, the flyby probe would pass by Alpha Centauri in 43 years, and we would receive the message back 4.3 years later. If we wanted the flyby probe to travel at 20 percent of the speed of light (twice as fast), the size of the vehicle would not change. We would still have four tons of working fluid to push a one-ton space vehicle, but now we would have to put 72 kilograms (160 pounds) or four times as much antimatter in the smaller fuel tank. This mass is still negligible compared to the five-ton mass of the spacecraft plus working fluid. At 20 percent of the speed of light, the flyby probe would reach Alpha Centauri in 21.5 years and we would get the information back 4.3 years later, 25.8 years after the start of the mission.

Suppose we wanted to have that one-ton spacecraft slow down and rendezvous at Alpha Centauri in 25 years. Then we would need to use a two-stage space vehicle. The second state would be the rendezvous stage, and would be the same size as the first stage. Since the effective "payload" for the booster stage is now five tons instead of one, the amount of working fluid needed for reaction mass in the booster stage is also five times larger, or 20 tons. The amount of mirror matter needed is still small in mass compared to the 25-ton mass of the rest of the vehicle: 350 kilograms, or 770 pounds.

The 350 kilograms of mirror matter would heat up the 20 tons of working fluid and accelerate the rendezvous stage up to 20 percent of the speed of light in less than a year. After coasting at 0.2 c for 21.5 years, the spacecraft would approach Alpha Centauri. The rendezvous stage would then turn around, start up its engine, and use its four tons of reaction mass, heated by 72 kilograms of mirror matter, to bring the one-ton rendezvous spacecraft to a halt in the Alpha Centauri system. Mirror matter rockets *can* take us to the stars—if we develop the necessary technology.

Mirror Matter!

The vision of the Star Trek television shows and movies is an alluring one. But it remains science fiction. Even mirror matter would be inadequate for Star Trek-type performance. Running around the Milky Way galaxy doing several missions with one vehicle would mean high mass ratios—and *a lot* of mirror matter. But if humanity can find a way to *make* a lot of mirror matter, then interstellar travel becomes a reality.

In the coming decades we will see the production and storage of significant quantities of mirror matter. The first uses for space travel will be within the solar system. And if no other propulsion system proves to be better, and we choose to spend the time to collect the solar energy to generate the kilograms of mirror matter needed, then another fantasy will become fact. One of these days we can ride to the stars on a jet of annihilated matter and mirror matter.

And then James Kirk and Montgomery Scott, Pavel Chekov and Janice Rand, Leonard McCoy and Hikaru Sulu will no longer be science fiction.

They will be us.

Turn Left at the Moon: VIII

The last stops of the tour were anticlimactic for Eric. They switched time zones and headed down the other side of Mars, and made an overnight stop at the Sagan Memorial/Viking 2 Lander site on the Plains of Utopia. Then came a night at the base of Mount Elysium and a long cable car trip the next day to the top of the ten-kilometer volcano to look around.

The last stop was at the bottom of the Hellas Basin, the lowest point on Mars. At four kilometers below the average surface height of Mars (the tour guide kept calling it "sea level," which irritated Ginny), the air there was much thicker. Hellas Basin was also the site of the other major martian spaceport. The vacation was nearing an end.

Eric was ready to go home. He knew what had happened to him at Boreal Camp. "It's an epiphany experience, Eric," his grandmother had told him that night. "Not everyone gets that kind of sudden flash of understanding what they want to do with their lives. You're lucky."

"It's like what happened when Mom met Dad," he had replied.

"Yes," Ginny had said, and smiled at some old memory of her own. "Yes, it's a lot like that."

Jim mistook his son's silence for sadness at leaving. "Hey, kid, maybe you'll come here some day again by yourself. Haven't you always said you wanted to be the pilot of an aerospace plane? Maybe your first command will be a Marsclimber instead of an Earth Spaceclimber."

"I'll be back, all right," Eric replied. "You can count on it."

The vertical ascent on the Marsclimber went much faster than the leisurely descent two weeks earlier. They were back on Phobos in a few hours. During their stay on Mars, the Longliner had returned to Earth, picked up a new load of tourists, and was now again berthed at Singer Station. Eric and Ginny took the inner cabin this time, and his parents got the outer stateroom with the viewport.

They were two days out from Mars when the big moment of the decade rolled around. Eric's mother had known about it already. It was the launch of the first interstellar probe, to Alpha Centauri. Her team had worked on the rocket engine for the robot probe.

"It was one of those coincidences that happens only rarely on a production line, even one as precision-oriented as ours," Kirsten said that evening to the captain at dinner. "I'd heard rumors about engine 33 even before it came to my team for inspection. Every part for every pump, actuator, and subassembly had exactly the right tolerance. Every part performed at the top of its range. The assemblers began talking about 33 and taking special care of it when they worked on some part of it.

"By the time it got to us for final flaw inspection I couldn't find a single flaw anywhere. Totally clean. And in the final test stand firings 33 set a record for efficiency and thrust that's still unequaled at Turin—or anywhere else, I think.

"Well, we figured we had one special engine, and decided not to deliver it to Fed Express. We wanted to keep it for something special."

"And this is special."

"Do you mean the interstellar probe is using the same kind of rocket engine used by the overnight express rockets?" the captain asked in surprise.

"Yes. I mean, the probe will accelerate at ten gees, just like the overnight express. They're antimatter rockets, so of course the mass ratios are the same—four tons of propellant mass for every ton of vehicle. There's one major difference. The overnight expresses pull ten gees for one day. Centaurus I will pull ten gees for eighteen days. That multiplies out to half the speed of light."

"Amazing."

"Well, there are a few other differences.—Eric, the broccoli.—The fuel tanks on Centaurus I hold liquid hydrogen instead of methane, to improve the exhaust velocity. And it's not carrying 300 grams of antimatter, but 300 kilos. The combustion chamber will run at a higher temperature, but the superconducting coils on 33 always gave us a magnetic bottle that was stronger than anything we'd ever seen."

"And cooling?" asked one of the other passengers.

"The best we ever tested. And I hear the Aussies have added some interesting innovations for backup cooling, too.

"I think engine 33 can pull it off."

The ship's scotty had found the coordinates for the launch point of the interstellar probe. The liner's longest-focus telecam was pointed in that direction. The probe was floating at a safe distance from Feynman City near the Earth-Moon L-5 point. Half the dining room's monitors showed the countdown being broadcast "live" from Earth. "Live," in this case, meant a six-minute light-time delay for the radio waves to get from Earth to the Longliner still near Mars. The other monitors held an image of the long lines of the probe. "Centaurus I" was written on its body. Visible on the engine nozzle was the number "33."

"The launch has already taken

place by now," said Eric, looking at his watch.

Five minutes later half the monitors lit up with a glare as the ship's telecam caught the light from the probe's engines near distant Earth. A bright streak began moving. The other screens with the closeup of the probe showed nothing. Then two seconds later the number "33" puffed into ash. The liquid drop radiators turned into bright red feathers.

"We have ignition," came the bored-sounding voice from Earth launch control. "Ignition on Centaurus I . . . systems nominal . . . "

Eric looked over at his mother. "Go," she was muttering. "Go. Go. Go." After a long minute she suddenly relaxed. She leaned into her husband's shoulder.

"If it was gonna blow, it would have done it by now," she said. "We're on our way."

"There it goes!" said the captain. The bright streak on the monitors faded into a tiny white light.

"That's my engine out there," said Eric's mother.

Eric's father shook his head in wonderment. "From a white streak in a bubble chamber to a white streak in interstellar space in just a hundred years. Not bad. Not bad at all."

Eric's grandmother looked at Eric from across the table and smiled. "You're next," she whispered.

Slowly, Eric grinned.

Appendix

Abbreviations and Prefixes for Numbers

a	atto-	= quintillionth	= 0.000000000000000001	= 10^{-18}
f	femto-	= quadrillionth	= 0.000000000000001	= 10^{-15}
p	pico-	= million millionth	= 0.000000000001	= 10^{-12}
n	nano-	= thousand millionth	= 0.000000001	= 10^{-9}
u	micro-	= millionth	= 0.000001	= 10^{-6}
m	milli-	= thousandth	= 0.001	= 10^{-3}
k	kilo-	= thousand	= 1000.0	= 10^{3}
M	mega-	= million	= 1000000.0	= 10^{6}
G	giga-	= thousand million (billion)	= 1000000000.0	= 10^{9}
T	tera-	= million million (trillion)	= 10000000000000.0	= 10^{12}
p	peta-	= thousand million million	= 1000000000000000.0	= 10^{15}
E	exa-	= million million million	= 1000000000000000000.0	= 10^{18}

Conversion Factors

mm = millimeter   kg = kilogram   F = degrees Fahrenheit
cm = centimeter   ml = milliliter   K = degrees Kelvin
m = meter   l = liter   s = second
km = kilometer   C = degrees Celsius   h = hour
g = gram

Length

1 in	= 2.54 cm		1 cm	= 0.39 in
1 ft	= 30.5 cm or 0.305 m		1 m	= 3.28 ft or 1.09 yd
1 yd	= 91.44 cm or 0.9144 m			
1 mi	= 1.6 km		1 km	= 0.62 mi
25,000 mi	= 40,000 km		1,000 km	= 621 mi
	(circumference of the Earth)			

Mass

1 oz = 28.3 g   1 g   = 0.04 oz
1 lb = 0.45 kg or 454 g   1 kg = 35 oz or 2.2 lb

Volume

1 oz = 30 ml   1 ml = 0.034 liquid oz
1 qt = 0.95 l   10 ml = 0.34 liquid oz
1 gal = 3.8 l   1 l   = 34 oz or 1.06 liquid qt
                                              or 0.26 gal

Temperature

$C = (F - 32)/1.8$ 0 C = +32° F   100° C = 212°F
$F = 1.8C + 32$ 0 F = −17.78°C   212° F = 100° C
$K = C + 273.15$ or   0 K = −459.67° F
$\qquad (F + 459.67)/1.8$ or −273.15° C
$C = K - 273.15$
$F = K + 459.67$

Bibliography

A good source for the most recent information on mirror matter research and technology is the *Mirror Matter Newsletter*. For more information, send your name, address, and a short statement of the reasons for your interest to:

> Mirror Matter Newsletter
> Robert L. Forward, editor
> P.O. Box 2783
> Malibu, California 90265-7783
> USA

Two other basic overviews of mirror matter space propulsion are:

Forward, Robert L. *Alternate Propulsion Energy Sources*. Air Force Rocket Propulsion Laboratory, Report TR-83-067, October 1983.

Forward, Robert L. *Antiproton Annihilation Propulsion*. Air Force Rocket Propulsion Laboratory, Report TR-85-034, September 1985.

These two reports are specificially mentioned in the bibliography under Chapter 5 and Chapter 6. However, the information they contain is also significant for nearly all of this book.

Chapter 1

Alfven, Hannes. *Worlds-Antiworlds*. New York: W.H. Freeman, 1966.

Davies, Paul. *God and the New Physics*. New York: Simon & Schuster, 1983.

Fritzsch, Harald. *The Creation of Matter*. New York: Basic Books, 1984.

Gribben, John. *In Search of the Big Bang*. New York: Bantam, 1986.

Halzen, Francis and Alan D. Martin. *Quarks and Leptons*. New York: John Wiley & Sons, 1984.

Polkinghorne, John. *The Particle Play*. New York: W.H. Freeman, 1980.

Quigg, Chris. "Elementary Particles and Forces". *Scientific American* April 1985, pp 84-95.

Sambursky, Shmuel, ed. *Physical Thought from the Presocratics to the Quantum Physicists*. New York: Pica Press, 1975.

Sternheim, Morton M. and Joseph W. Kane. *General Physics*. New York: John Wiley & Sons, 1986.

Chapter 2

"The Antineutron." *Scientific American* November 1956: 64, 66.
"Anti-Neutron Discovered." *Science News Letter* September 22, 1956: 183.
"Anti-Proton Discovered." *Science News Letter* October 29, 1955: 275.
"Discovery of the Antiproton." *Science* December 23, 1955: 1222.
"The Filled-Out Universe." *Time* September 24, 1956: 39.
Forward, Robert L. "Production of Heavy Antinuclei: Review of Experimental Results." Proceedings of Hydrogen Cluster Ion Workshop, Menlo Park, CA, January 1987.
Lederman, Leon M. "Observations of Antideuterons." Press release, American Institute of Physics, June 14, 1965.
"New Type of Neutrino Discovered in Columbia-Brookhaven Experiment." Press release Brookhaven National Laboratory, July 1, 1962.
"Nobel Prize for Physics in 1959: Dr. Emilio Segrè and Dr. Owen Chamberlain." *Nature* October 17, 1959: 1189.
Ritter, Jim. "Fermilab's King of Quarks." *Living Magazine: Chicago Sun-Times* June 22, 1986: 8–9.
Segrè, Emilio and Clyde E. Wiegand. "The Antiproton." *Scientific American* June 1956: 37–41.
"Size of the Antiproton." *Science* February 24, 1956: 318.
Yarris, Lynn. "The Nobelists." *LBL Research Review* Winter 1986: 6–13.

Chapter 3

Anderson, Poul. *Ensign Flandry.* Philadelphia: Chilton Books, 1966.
_____. *Orbit Unlimited.* New York: Pyramid Books, 1961.
_____. *Tau Zero.* New York: Doubleday, 1970.
Anthony, Piers. *Macroscope.* New York: Avon, 1969.
Asimov, Isaac. *The Foundation Trilogy.* New York: Doubleday, 1982.
_____. *Foundation's Edge.* New York: Doubleday, 1982.
_____. *Foundation and Earth.* New York: Doubleday, 1986.
_____. *I, Robot.* New York: Grove Press, 1950.
Blish, James. "Beep!" *Galaxy* February 1954.
_____. *Spock Must Die!* Bantam, 1970.
Brown, Frederic. "Placet is a Crasy Place." *Astounding Science Fiction* May 1946.
Clarke, Arthur C. *The Wind From the Sun: Stories of the Space Age.* San Diego: Harcourt Brace Jovanovich, 1972.
Davies, Paul. *Fireball.* London: Heineman, 1987.
Forward, Robert L. *The Flight of the Dragonfly.* New York: Pocket Books/Timescape, 1984.
Hamilton, Edmond. *Star Kings.* 1950.
Heinlein, Robert A. *Orphans of the Sky.* New York: Putnam, 1963.
Leinster, Murray. *Talents, Incorporated.* 1962.
Manning, Lawrence. "The Living Galaxy" *Wonder Stories.* Sept. 1934.

Niven, Larry. *Neutron Star*. New York: Ballantine, 1968.

———— and Jerry Pournelle. *The Mote in God's Eye*. New York: Simon & Schuster, 1974.

Preuss, Paul. *Broken Symmetries*. New York: Pocket Books, 1983.

Shaw, Bob. *A Wreath of Stars*. New York: Doubleday, 1977.

Simak, Clifford. *Way Station*. New York: Doubleday, 1963.

Smith, Cordwainer. "The Lady Who Sailed the Soul." *Galaxy* April 1960.

Van Vogt, A.E. "Far Centaurus." *Astounding Science Fiction* January 1944.

————. "Storm." *Astounding Science Fiction* October 1943.

Wilcox, Don. "The Voyage That Lasted 600 Years." *Amazing* October 1940.

Williamson, Jack. *SeeTee*. New York: Jove Publications, 1979.

Chapter 4

Cramer, John G. and Wilfred J. Braithwaite. "Distinguishing Between Stars and Galaxies Composed of Matter and Antimatter Using Photon Helicity Detection." *Physical Review Letters* October 24, 1977: 1104–7.

Davies, Paul. *Superforce*. New York: Simon & Schuster, 1984.

Fisher, Arthur. "New Ferment in the Mirror World of Antimatter/Antigravity." *Popular Science* July 1986: 54ff.

"Galactic Positronium Mystery Deepens." *Science News* July 19, 1986: 130.

Gribben, John. *In Search of the Big Bang*. Bantam, 1986.

Maddox, John. "New Ways with Matter/Antimatter." *Nature*, March 26, 1987: 327.

Pagels, Heinz R. *Perfect Symmetry*. New York: Simon & Schuster, 1985.

Trefil, James S. *The Moment of Creation*. New York: Scribner, 1983.

Chapter 5

Broad, William J. "Atom-Smashing Now and in the Future: A New Era Begins." *New York Times* February 3, 1987: C1–C2.

Forward, Robert L. *Alternate Propulsion Energy Sources*. Air Force Rocket Propulsion Laboratory, Report TR–83–067, October 1983.

————. *Antiproton Annihilation Propulsion*. Air Force Rocket Propulsion Laboratory, Report TR–85–034, September 1985.

Chapter 6

Augenstein, Bruno. "Concepts, Problems and Opportunities for Use of Annihilation Energy: An Annotated Briefing on Near-Term RDT&E to Assess Feasibility". Rand Note N-2302-AF/RC. Rand Corp. June 1985.

Dvorak, Wes. "AT&T Scientists Trap Atoms in Light Bottle." Press release, AT&T Bell Laboratories, July 3, 1986.

Fisher, Arthur. "New Ferment in the Mirror World of Antimatter/Antigravity." *Popular Science* July 1986: 54ff.

Forward, Robert L. *Alternate Propulsion Energy Sources*. Air Force Rocket Propulsion Laboratory, Report TR–83–067, October 1983.

―――. *Antiproton Annihilation Propulsion*. Air Force Rocket Propulsion Laboratory, Report TR–85–034, September 1985.

Schwarzschild, Bertram. "Laser Beam Focus Forms Optical Trap for Neutral Atoms." *Physics Today* September 1986: 17–19.

―――. "Now They're Even Trapping Antiprotons." *Physics Today* September 1986: 19.

Thomsen, Dietrick. "Trapping Antimatter, Antiprotons on Hold." *Science News* November 29, 1986: 340–41.

Chapter 7

Augenstein, Bruno. "Some Examples of Propulsion Applications Using Antimatter." Rand Paper P-7113. Rand Corp. July 1985.

Burlage, H., Jr., *et al.* " Unmanned Planetary Spacecraft Chemical Rocket Propulsion." *Journal of Spacecraft and Rockets* October 1972: 729–37.

Ehricke, Krafft A. "Spacecraft Propulsion." *McGraw-Hill Encyclopedia of Science and Technology*, 1982.

Forward, Robert L. "Antimatter Propulsion." *Journal of the British Interplanetary Society*, 1982: 391–95.

―――. "Antiproton Annihilation Propulsion." *Journal of Propulsion* September–October 1985;370–74.

―――., *et al.* Cost Comparison of Chemical and Antihydrogen Propulsion Systems for High Delta-V Missions." *American Institute of Aeronautics and Astronautics*, Paper 85–1455, July 1985.

Grey, Jerry. "Propulsion." *McGraw-Hill Encyclopedia of Science and Technology*, 1982.

Hibbard, Robert R. "Specific impulse." *McGraw-Hill Encyclopedia of Science and Technology*, 1982.

Howe, Steven D. "Antimatter propulsion: Feasibility, Status and Possible Enhancement." Paper, Manned Mars Mission Workshop, June 10, 1985.

Morgan, D.L. *Annihilation of Antiprotons in Heavy Nuclei*. Report TR–86–011, Air Force Rocket Propulsion Laboratory. April 1986.

―――. *Antiproton-hydrogen atom annihilation*. Report TR–86–019, Air Force Rocket Propulsion Laboratory. May 1986.

―――. "Physics of Antimatter-Matter Reactions for Interstellar Propulsion." Paper, AAAS Annual Meeting, Philadelphia, August 22, 1986.

Sutton, George P. "Rocket Engine." *McGraw-Hill Encyclopedia of Science and Technology*, 1982.

Vulpetti, G., "A Propulsion-Oriented Synthesis of the Antiproton-Nucleon Annihilation: Experimental Results." *Journal of the British Interplanetary Society* March 1984: 124–34.

Chapter 8

Augenstein, Bruno. Letter to Dr. Peter Hammerling, La Jolla Institute, 1985. Unpublished.

Broad, William J. "Search for Antimatter is Showing Results." *New York Times* January 20, 1987: C1, C3.

Gray, L. and T. E. Kalogeropoulos. "Possible Biomedical Applications of Antiproton Beams: Focused Radiation Transfer." *Radiation Research* 1984, 246–52.

_____."Possible Bio-medical Applications of Antiprotons." I. In-vivo Direct Density Measurements: Radiography." *IEEE Transactions in Nuclear Science* April 1982: 1051–57.

Hynes, M.V. *Physics with Low-Temperature Antiprotons.* Report, Los Alamos National Laboratory. 1985.

"Professor Proposes Cancer Treatment." *Syracuse University Magazine* November 1985: 8–9.

Rogers, J. "Measuring the Gravitational Properties of Antihydrogen." Paper, University of Rochester, Dept. of Physics, 1986.

Schechter, Bruce. "Elementary-Particle Physics." *Physics Today* April 1986: 28–29.

Chapter 9

Air Force Rocket Propulsion Laboratory. Pamphlet, U.S. Air Force, 1986. n.d.

Augenstein, Bruno, et al. "Proof-of-Concept Demonstration of Antiproton Propulsion." Report, RAND Corp. 1986.

Forward Robert L., et al. "Cost Comparison of Chemical and Antihydrogen Propulsion Systems for High Delta-v Missions." *American Institute of Aeronautics and Astronautics* July 1985.

Greeley, Brendan M., Jr. "NASA/Defense Dept. Study Examines Launcher Concepts." *Aviation Week and Space Technology* June 30, 1986: 22–23.

Nordley, Maj. Gerald D. Air Force Antimatter Technology Program. Reprint. U.S. Air Force Astronautics Laboratory September 1987.

Tucker, Jonathan B. "U.S. Revives Space Nuclear Power: Orbiting Reactors Could Fuel Space Stations, Satellites and Laser Weapons." *High Technology* August 1984: 15–19.

Chapter 10

Albee, A.L. "Future Mars Exploration." *Journal of the British Interplanetary Society* August 1984: 370–74.

Burlage, Jr. H., et al. "Unmanned Planetary Spacecraft Chemical Rocket Propulsion." *Journal of Spacecraft and Rockets* October 1972: 729–37.

Caluori, Vincent A. "Technology requirements for Future Space Transportation Systems." Paper 79–0867, American Institute of Astronautics and Aeronautics. May 1979.

Cooper, L.P. et al. "Status of Advanced Propulsion for Space-Based Orbital Transfer Vehicle." Technical Memorandum 88848, National Aeronautics and Space Administration. 1986.

Covault, Craig. "Manned U.S. Lunar Station Wins Support." *Aviation Week and Space Technology* November 19, 1984: 73–84.

_____. "Ride Panel Calls for Aggressive Action to Assert U.S. Leadership in Space." *Aviation Week and Space Technology* August 24, 1987: 26–27.

_____. "Ride Panel Will Urge Lunar Bases, Earth Science as New Space Goals." *Aviation Week and Space Technology* July 13, 1987: 16–18.

_____."White House, NASA Action to Regain U.S. Space Leadership Role." *Aviation Week and Space Technology*. August 10, 1987: 18–19.

CRAF: Comet Rendezvous and Asteroid Flyby. JPL Factsheet. 1985.

Disher, John. "Next Steps in Space Transportation and Operations." *Astronautics & Aeronautics*, January 1978: 22–30.

Draper Ronald F. "Affordable Solar System Exploration." *Aerospace America* October 1986: 28–31.

_____ and S.A. Lundy. "Comet Rendezvous." *Journal of the British Interplanetary Society* August 1984: 361–65.

Duxbury, John H. and Gary M. Paul. "Interplanetary Spacecraft Design Using Solar Electric Propulsion." *Journal of Spacecraft and Rockets* February 1976: 99–105.

Ehricke, Krafft A. "Ion Propulsion." *McGraw-Hill Encyclopedia of Science and Technology*, 1982.

Feingold, H. "Comet Nucleus Sample Return." *Journal of the British Interplanetary Society* August 1984: 388–94.

Forward, Robert L., et al. "Cost Comparison of Chemical and Antihydrogen Propulsion Systems for High Delta-v Missions." *American Institute of Aeronautics and Astronautics*, Paper 84–1455, July 1985.

Friedman, Louis D. "Solar Sailing Development Program (Fiscal Year 1977)." Final Report. Vol. 1 NASA/Jet Propulsion Laboratory, January 30, 1978.

Friedman, Louis D., et al. "Solar Sailing—The Concept Made Realistic." Paper, 16th Aerospace Sciences Meeting, AIAA, January 18, 1978.

Hardy, Terry L. et al. Electric Propulsion Options for the SP-100 Reference Mission. Technical memorandum 88918. National Aeronautics and Space Administration. 1987.

"Ion Propulsion for Spacecraft." NASA/Lewis Research Center, 1977.

Jaffe, Leonard D., et al. "An Interstellar Precursor Mission." *Journal of the British Interplanetary Society*. June 1980: 3–26.

_____ and Harry N. Norton. "A Prelude to Interstellar Flight." *Astronautics & Aeronautics* January 1980: 38–44.

Jones, R.M. and C.G. Sauer. "Nuclear Electric Propulsion Missions." *Journal of the British Interplanetary Society* August 1984: 395–400.

Martin, James A., et al. "Propulsion Evaluation for Orbit-on-Demand Vehicles. *Journal of Spacecraft and Rockets* November 1986: 612–19.

McLaughlin, William I. and J.E. Randolph. "Starprobe: To Confront the Sun." *Journal of the British Interplanetary Society* August 1984: 375–80.

Moeckel, Wolfgang E. "Comparison of Advanced Propulsion Concepts for Deep Space." *Journal of Spacecraft and Rockets* December 1972: 863–68.

_____. Propulsion Systems for Manned Exploration of the Solar System." *Astronautics & Aeronautics* August 1969: 66–77.

Niehoff, John C. and Alan L. Friedlander. "Comparison of Advanced Propulsion Capabilities for Future Planetary Missions." *Journal of Spacecraft and Rockets* August 1974: 566–73.

Page, Russell J., *et al.* A Design Study of Hydrazine and Biowaste Resistojets. Final Report. Contract Report 179510. National Aeronautics and Space Administration. September 1986.

Stoker, Carol and Christopher P. McKay. "Mission to Mars: The Case for a Settlement." *Technology Review* November 1983: 27–37.

Swenson, B.L. "Titan Atmospheric Probe." *Journal of the British Interplanetary Society* August 1984: 366–69.

TAU project. Press release. Jet Propulsion Laboratory. 1986.

Voulelikas, G.D. Electric Propulsion: A Review of Future Space Propulsion Technology. Report. Communications Research Centre (Canada). October 1985.

Wallace, R.A., *et al.* "Interplanetary Missions for the Late Twentieth Century." *Journal of Spacecraft and Rockets* May 1985: 316–24.

Chapter 11

Brin, David. *The Postman.* New York: Bantam, 1985.

Cassenti, Brice N. "Design Considerations for Relativistic Antimatter Rockets." *Journal of the British Interplanetary Society,* 1982: 396–404.

Clarke, Arthur C. "Counting Down for a Flight to the Stars." *Discover* July 1985: 54–58.

Dyson, Freeman J. "Interstellar Transport." *Physics Today* October 1968: 41–45.

Forward, Robert L. "Antimatter Propulsion." *Journal of the British Interplanetary Society* 1982: 391–95.

_____. "Antiproton Annihilation Propulsion." *Journal of Propulsion* September–October 1985:370–74.

_____. "Feasibility of Interstellar Travel: A Review." Paper, *International Astronomical Association,* 1985.

_____. "A Programme for Interstellar Exploration." *Journal of the British Interplanetary Society* 1976: 611–32.

_____. "Roundtrip Interstellar Travel Using Laser-pushed Lightsails." *Journal of Spacecraft and Rockets* March 1984: 187–95.

_____. "Starwisp: an Ultra-light Interstellar Probe." *Journal of Spacecraft and Rockets* May 1985: 345–50.

Freitas, Robert A. Jr., "Interstellar Probes: A New Approach to SETI." *Journal of the British Interplanetary Society* 1980: 95–100.

Harris, M.J. "On the Detectability of Antimatter Propulsion Spacecraft." *Astrophysics and Space Science* 123 (June 1986): 297–303.

Harrison, Harry. *West of Eden.* New York: Bantam, 1985

_____. *Winter in Eden.* New York: Bantam, 1986.

Heppenheimer, T.A. "On the Infeasibility of Interstellar Ramjets." *Journal of the British Interplanetary Society,* 1978: 222–24.

Jackson, A.A. "Some Considerations on the Antimatter and Fusion Ram Augmented Interstellar rocket." *Journal of the British Interplanetary Society* 1980: 117–20.

Jones, Lee W. and Dennis R. Keefer. "NASA's Laser-Propulsion Project." *Astronautics & Aeronautics* September 1982: 66–73.

Mallove, Eugene F. "Starflight: The Ultimate Voyage." *Technology Review* January 1984: 60–61.

————, Robert L. Forward, Zbigniew Paprotry, and Jurgen Lehmann. "Interstellar Travel and Communication: A Bibliography." *Journal of the British Interplanetary Society* July 1980:201–48.

Matloff, Gregory L. "The Interstellar Ramjet Acceleration Runway." *Journal of the British Interplanetary Society* 1979: 219–20.

————. and Eugene F. Mallove. "The First Interstellar Colonization Mission." *Journal of the British Interplanetary Society* 1980: 84–88.

Niven, Larry. *Protector*. New York: Ballantine, 1973.

————. *Tales of Known Space*. New York: Ballantine, 1975.

————. *World of Ptavvs*. New York: Ballantine, 1966.

———— and Jerry Pournelle. *Lucifer's Hammer*. New York: Playboy Press, 1977.

Nordley, Gerald D. Application of Antimatter-electric Power to Interstellar Propulsion. Paper. International Astronautical Federation, 38th Congress. October 1987.

Paprotny, Zbigniew and Jurgen Lehmann. "Interstellar Travel and Communication Bibliography: 1982 Update." *Journal of the British Interplanetary Society* 1983: 311–29.

————, et al. "Interstellar Travel and Communication Bibliography: 1984 Update." *Journal of the British Interplanetary Society* 1984: 502–12.

Powers, Robert M. "Propulsion Future." *Space World* September 1985: 17–19.

Sheffield, Charles. *Between the Strokes of Night*. New York: Simon & Schuster, 1985.

"Star Probe Study Effort Concluded by British Interplanetary Society." *Spaceworld* June 1979: 25–27.

Viewing, D.R.J., et al. "Detection of Starships." *Journal of the British Interplanetary Society* 1977: 99–104.

Whitmire, Daniel P. and A.A. Jackson. "Laser Powered Interstellar Ramjet." *Journal of the British Interplanetary Society* 1977: 223–26.

Index

257